SPACE ERA STRATEGY

The Way China Beats The U.S.

I0464221

CHAN KAI YEE

Space Era Strategy

Available from Amazon.com, CreateSpace.com, and other retail outlets

Because of the dynamic nature of the Internet, any web addresses or links contained in this book may have changed since publication and may no longer be valid.

The views expressed in this work are solely those of the author and do not necessarily reflect the views of Amazon.com, CreateSpace.com.

ISBN-13: 978-1500875664
ISBN-10: 150087566X

Printed by CreateSpace, An Amazon.com Company

b

TO ELEANOR AND RACHEL

Preface

U.S. Must Not Be Beaten by China!

Shang Yang's reform more than 2,000 years ago enabled the State of Qin to conquer all other states and unify China.

China's reform now is similar to Shang Yang's in giving play to people's talents and diligence. Then will it make China so strong as to conquer the world?

Germany tried to conquer the world when it became strong but suffered disastrous defeat. Will China be so stupid?

Chinese scholars believe that of all the ancient civilizations, Chinese civilization is the only one that has survived for more than 4,000 years because China has never tried to conquer the world or been fond of war.

However, a despot did emerge in China who wanted to conquer the world with his ideology called Mao Zedong Thought. The Thought has caused millions of death in the Killing Field in Cambodia, but fortunately, China is poor and backward at that time. Otherwise, the human race will suffer more serious disasters than World War II as Mao was even willing to fight a nuclear war for his ideal of communism.

As China's political system is unable to prevent the emergence of a despot like Mao as its leader, the U.S. must remain at least a rival to China so that there is the balance of strength to maintain peace.

Is the U.S. prepared for that?

China Will Beat the U.S.

> Subduing the enemy by stratagem is the best;
> by diplomacy, next best;
> by fighting in the field, third alternative;
> by attacking cities, last resort.

--The Art of War by Sun Tzu

U.S. Will Be Beaten First by Its Own Outdated Air-Sea Battle Stratagem

There is nothing new in Air-Sea Battle. It was the stratagem the U.S. adopted in defeating Japan in the 1940s. Such an outdated stratagem is certainly much inferior to China's Stratagem of Integrated Space & Air Capabilities in our space era. The picture on the cover of this book describes how a whole U.S. aircraft carrier battle group is destroyed by the missiles of one space-air bomber.

China Beats the U.S. by Diplomacy in Establishing an Alliance with Russia

U.S. talented politician Henry Kissinger wants U.S.-Chinese and U.S.-Russian relations to be better than Sino-Russian relations. Obama has got the contrary. His efforts to contain China by his pivot to Asia have driven China to Russia's side.

Russia and China are now close allies in countering the U.S.

China Is Surpassing the U.S. in Weapon Development

Now, China has surpassed the U.S. in anti-satellite (ASAT), anti-ASAT, hypersonic glide vehicle, drone, electromagnetic gun, amphibious landing craft, etc. and is catching up and will soon surpass the U.S. in satellite global positioning system, midcourse ICBM interception, aerospace plane, conventional and nuclear submarines, stealth fighter jets, destroyers, frigates, ballistic and cruise missiles, etc.

The problem is American people even quite a few military experts are not well informed of China's fast military modernization so that the U.S. fails to make real efforts to improve its weapon development.

For example, in his recent article based on Pentagon information, Robert Haddick, an independent contractor at U.S. Special Operations Command, says, China's DF-21D aircraft carrier killer ballistic missile "still apparently not tested against a moving target at sea".

In fact, in 2011 and 2012, China conducted quite a few launches of DF-21D in the South China Sea and successfully hit and sank a simulated model of aircraft carrier made by transforming China's Yuanwang 4 survey ship. A U.S. research institute has learnt that, but Pentagon has not.

Related information is reflected in an article by Wang Genbin, deputy commander-in-chief of Department 4 of China Aerospace Science & Industry Corp. (CASIC), published on a journal publicly available in China.

Wang says in the article that in the two decades since 1988, China spent 3 billion yuan ($494 million) in successfully developing DF-21A, 21B, 21C and 21D missiles and completed the transition from development of only nuclear missiles to that of both nuclear and conventional missiles and from fixed target to low-speed target. In addition, the accuracy has been improved from several hundred to several tens of meters.

The U.S. Must be Warned

It has to give up its outdated stratagem and adopt a stratagem of the space era in order to counter China's.

It has to greatly improve its diplomacy to unite all the countries in being on guard against China in case unfortunately a despot like Mao emerges as Chinese leader.

The U.S. shall be well informed of the progress of China's military development so that it will have counter measures in time.

Chan Kai Yee
August 18, 2014

Table of Contents

PART 1　THE SPACE ERA

1. New Weapons and Strategy for New Era

A Future War

While the U.S. has built its new state-of-art Ford-class aircraft carriers for gradual replacement of old ones and improved the Aegis destroyers in its navy, China has been making slow progress in building its aircraft carriers. The three aircraft carriers it has are all conventional ones unable to protect its trade lifelines far away from home.

There has been speculation since 2014 that China will build at least six aircraft carriers mostly nuclear ones to protect its interests abroad, but more than a decade later such carriers have never even been planned. What does China have to defend its trade lifelines? It depends on U.S. navy's protection according to Chinese leaders. The U.S. remains world leader due to its powerful air force and navy though Chinese economy is bigger than the U.S.

They say, "One mountain is not big enough for two tigers"; therefore, there will be conflicts between China and the U.S. that may lead to war. As a result, China's trade lifelines will be in great danger.

The day of conflicts finally comes.

A war breaks out between China and Japan due to China's support for Liukiu (Okinawa) people's struggle for independence. As Japan is U.S. ally, the U.S. comes to help Japan. It sends two aircraft carrier battle groups to Japan, two to the Philippines and two to the Indian Ocean to block Malacca Strait.

Being engaged in war with Japan, China cannot spare any warships or aircrafts to deal with U.S. warships, but warns the U.S. it will sink U.S. warships if the U.S. does not lift the blockade.

The U.S. regards China's warning as mere bluffing. What does China have to sink U.S. warships? Chinese warships and aircrafts are busy engaging the Japanese while U.S. aircrafts dominating the sky are watching closely the movement of Chinese mobile missiles in order to destroy them before they are launched.

Attack from Space

Unexpectedly, American aircraft carriers are attacked by too many accurate hypersonic anti-warship missiles from land bases, warships and bombers for their air defense systems to intercept so that three aircraft carriers are crippled but not sunk.

U.S. headquarters realizes the danger near China and tell all U.S. navy to retreat out of the range of China's land-based anti-ship missiles. The three damaged carriers go to their bases to be repaired while the three intact and other five are organized into two groups to cut China's vital trade lifelines.

Having crippled Japanese air force and navy, Chinese navy went out to defend their trade lifelines. U.S. aircraft carrier battle groups are very happy that they will have the opportunity to wipe out smaller and inferior Chinese navy.

From a fleet of three aircraft carriers, more than 20 heavy stealth fighter jets took off to dominate the sky while other fighter jets and U.S. warships rush towards Chinese navy to send all Chinese warships and aircrafts down into the sea.

However, before they come near Chinese warships, lots of accurate missiles come from nowhere to destroy quite a few of them.

Where have those missiles come from?

U.S. spy satellites find that they came from China's huge manned aerospaceplanes (space-air aircrafts). Those aircrafts were launched by rockets into their orbits in space and can remain in orbit as space stations for China's lunar program. They get their supplies and replacement of crew from China's transport rockets or master space station. When they need maintenance, they fly to China's lunar base for that; therefore, though they are able to land on the ground, they seldom do so. As a result, the U.S. is not aware of their integrated space and air capabilities.

At wartime, a transport rocket rendezvous with an aerospace aircraft to send it a set of missiles. The aerospaceplane immediately glides down to attack the aircraft carrier battle group with the missiles. It flies at the hypersonic speed of Mach 22 so that it engages enemy warships and aircrafts within an hour while it takes days for U.S. fleets to engage Chinese fleets.

Having no weapons to deal with the hypersonic aerospaceplane and the hypersonic missiles it has fired, the U.S. has to call back its warships and aircrafts that have not yet been destroyed by the aerospaceplane and hold peace talks with China.

No, some readers might say that the U.S. can use its superior strategic weapons such as ICBMs, SLBMs and strategic stealth bombers to destroy China.

No way! Those aerospaceplanes can fly at hypersonic speed to intercept ICBMs and SLBMs and shoot down U.S. strategic stealth bombers. Like the China from 1840 to 1945 and Britain at the beginning of World War II, the U.S. will be defeated by China due to its failure to follow the trend of the times.

The Lesson China Learnt when It Failed to Follow the Trend of the Times

Since the industrial revolution, due to rapid progress in science and technology, the world has been changing fast, especially in the weapons used in war and along with it the organization of military force.

The invention of gunpowder gave rise to the invention of guns, not rifles or pistols at first but cannons. Improvement in cannons made their use vital in naval battles, resulting in improvement of naval warships and organization.

When Britain invaded China during the first Opium War in 1940, its military superiority lied in its navy due to the improved warships for extensive use of cannons. At that time, China was entirely able to resist British navy if Emperor Daoguang followed his only enlightened courtier Lin Zexu's advice to strengthen coastal defense by importing enough cannons.

The failure to follow international trend of technological and industrial development led to another miserable defeat two decades later in the second Opium War when China was overwhelmed by Britain and France armed with cannons plus rifles.

China seemed to have learnt from its defeat and begun to establish enterprises to produce advanced guns and import advanced warships.

However, first China's top leader Empress Dowager Cixi diverted the funds for further strengthening Chinese navy to the construction of the Summer Palace for her retirement.

Second, the improvement in weapons was not accompanied by corresponding organization required for an advanced navy of the time.

The defeat of Chinese navy in 1894 taught China a bitter lesson as the heavy war compensation of the first Sino-Japanese war and the later invasion by combined forces of eight powers made China very poor and backward.

That is China's lesson on the heavy costs in failing to follow the trend.

The Lessons of Britain's Setback at the Beginning of World War II:

At the end of World War I, two new weapons tank and aircraft were proved very powerful in war.

Germany followed the trend to build up a powerful air force with thousands of aircrafts while Britain, though well informed of Germany's efforts, was inactive in developing its air force or mechanizing its army.

The heavy air raid suffered by Britain in the early years of the war was a bitter lesson for the whole world to learn from.

On the other hand, the use of tanks enabled Germany to make its mechanized troops almost invincible in Europe.

The Even More Rapidly Changing World Now

The rapid development of science and technology has brought us a rapidly changing world. As a result following the trend is now vital for the survival not only of a nation but also an enterprise no matter how strong the nation and enterprise is.

The above two lessons merely reflected people's obtuseness in following trend, but it is more important to be the first to exploit the most advanced science and technology of the time. That requires higher intelligence and resourcefulness.

The recent example was the IT warfare conducted by the U.S. in the Gulf War to cut enemy's communications and make enemy troops unable to receive commands and information.

However, this is the space era where IT technology has been mastered by lots of countries. However, space technology remains unaffordable for most countries due to the heavy costs. The United States has been the leader in space technology, but is quite in retard now due to its shrinking budget. That enables China to quickly catch up.

Air-Sea Battle Outdated

I was surprised when U.S. military invented its so-called Air-Sea Battle to deal with China for U.S. pivot to Asia, especially as such an out-dated strategy was regarded something new and powerful in space era.

Air-Sea Battle is the way the U.S. defeated Japan during World War II. At the end of war, the U.S. had wiped out Japanese navy and almost the entire Japanese air force. If it had kept on bombing Japanese troops and cities to destroy Japanese military assets and economy, it would have brought Japan on its knees even without using the atomic bombs. It would have taken a little more time. That proved the power of U.S. Air-Sea Battle.

The strategy is what the U.S. has adopted for decades. Certainly, the U.S. has been well experienced in it. The U.S. has already got all the necessary hardware, including B2B stealth bombers, F-22 fighter jets, huge fleets of aircraft carrier, Aegis destroyers and other advanced warships and submarines and the expertise to operate them skillfully. China has just begun to build up its navy. Though building warships at excessive speed, its fleet of only one aircraft carrier and less than 10

6

Aegis destroyers is no match to U.S. navy now.

Even at China's current high speed of warship building, it takes at least a decade for China to build a navy comparable to the U.S. in the area near China. However, even if China has the warships, China does not have the experience to operate them. Shall China compete with the U.S. in such a slow and ineffective way?

Certainly not.

Remember, when the world has entered the era of naval submarine, instead of making years of efforts to build a huge navy comparable to the British one, Germany built lots of submarines to cut Britain's transport lifelines.

The End of the Era of Aircraft Carrier

Recently, the Strategic Studies Institute (SSI) of U.S. Army War College published a long report titled "The Next Arms Race" edited by Henry D. Sokolsky.

Chapter 5 of the report titled "China and the Emerging Strategic Competition in Aerospace Power" is written by Mark Stokes and Ian Easton, responsible persons of the 2049 Project Research Institute. It describes the competition in aerospace power. According to the two writers, the competition "is being driven in large part by Chinese development of military capabilities and strategies, which increasingly challenge the ability of regional air-, missile-, and space-defense programs to keep pace."

The report is free to download at http://www.npolicy.org/userfiles/image/ub1113.pdf. Cyol.net, the website of China Youth Daily, provides a translation of the parts of the chapter selected by it.

The selection of the parts certainly reflects the website's views. Here, I give a summary of cyol.net's report mainly to reflect my views based on the Chapter.

The most interesting is the conclusion at the end of the chapter that due to China's vigorous arms race in developing the attack capabilities of its anti-ship ballistic and cruise missiles, a top U.S. analyst foresee a shift in U.S. basin— moving away from allied territories to Guam and the South Pacific Islands—and a greater U.S. naval presence in the Indian Ocean.

That is the end of aircraft carrier era. U.S. aircraft carriers will not be able to operate near a country with enough precision anti-ship ballistic and cruise missiles.

Let's see how the writers reach that conclusion:

The writers say in the chapter: China has kept on speedy development of its military modernization. It has started development, testing, and deployment of advanced aerospace capabilities, which have made its neighbors worry about their comparative weakness in the capabilities to access to and control of the air and space mediums in the event of a conflict. Due to the vast distances and long-time horizons in the Asia-Pacific region, access to and control of the air and space will be critical to achieving political and military successes on the land and the sea in the conflict.

The writers, like other American military experts, do not really see the tremendous potential of integrated space and air capabilities China has been developing. They talked much about the fast growth of China's ballistic and cruise missile capabilities, especially in relation to the safety of U.S. aircraft carrier battle groups. Ballistic missiles are to some extent related to space as they mostly cruise in space in midcourse, but cruise missiles have nothing to do with space.

However, as the writers' perspective is restricted to U.S. outdated strategy of Air-Sea Battle to counter rising Chinese military strength, they pay great attention to China's progress in anti-ship missile capabilities, especially China's anti-ship ballistic missiles, which, though will cruise in space in midcourse, remain ground-based; therefore, in strict sense, they are not aerospace capabilities.

The writers then discussed China's quick growth in anti-satellite (ASAT) capabilities and reconnaissance satellites, which are really relevant to aerospace capabilities. However, they view such capabilities merely from the point of view of the capabilities' roles in their outdated Air-Sea Battle.

That mainly concerned the ability to destroy, blind or otherwise neutralize enemy satellites, including in addition to weapons to hit enemy satellite, parasite satellites, high-energy microwave and particle beam weapons, electronic jammers and the cyber attack capabilities against enemy satellite tracking and control stations.

When the report discusses space-based surveillance, it seems that the writers will really describe China's integrated space and air capabilities. However, I am disappointed that they again talk about reconnaissance satellites and land-based long-range radar such as over-the-horizon backscatter and skywave over-the-horizon radar systems. The latter may extend its range possibly to 4,000 km.

Persistent Near-space surveillance! They finally reach their topic—aerospace capabilities. They say "Chinese analysts view the realm between the atmosphere and space—"near-space"—as an area of future strategic competition. Over the

8

decade, near-space flight vehicles may emerge as a dominant platform for a persistent region-wide surveillance capability during crisis situations. 'Near-space' is generally characterized as the region between 20 and 100 km (65,000 to 328,000 feet) above the earth's surface."

They understand that in spite of technical challenge, China is much interested in near space flight vehicles for their capabilities of reconnaissance, communications relay, electronic countermeasures and precision strike.

However, they do not make clear whether the precision strike operations come from the near space flight vehicles or such vehicles merely provide data and guidance for missiles launched from other platforms. I believe the attack will come from near-space flight vehicles, i.e. the vehicles flying between space and atmosphere. That is why the integrated space and air capabilities China has been developing is regarded by China as capabilities for both attack and defense.

If a missile similar to China's DF-21D aircraft carrier killer missile is launched from a near-space vehicle, what will be the consequence? The technology of terminal hit has already been there in the DF-21D, but there will be much better direct guidance from the near-space vehicle and the missile will be much quicker due to the hypersonic speed of the vehicle. The target aircraft carrier will have no escape.

In light of the expanding long-range precision strike capabilities of Chinese anti-ship ballistic and cruise missiles and the vigor of China's arms race, the writers conclude the chapter as described above.

True, the gunboat era has long been the past. In missile era, as missiles are fierce as attack weapons and very difficult to defend, a large aircraft carrier constitutes an easy huge target moving at low speed for missile attack. To defend, the carrier has to hit missiles small in size and moving at high speed. The odds are obviously on missiles.

Moreover, a state-of-the-art aircraft carrier costs more than $10 billion to build, while the state-of-the-art ballistic and cruise missiles used for saturate attack at the aircraft carrier cost much less. That enables a country with much less financial resources but the necessary technology to resist the attack by the aircraft carriers of a much richer country.

Now, aircraft carriers are not able to control the sea near a country with adequate missile capabilities. That is a clear sign of the end of aircraft carrier era.

However, they are useless before China's saturate attack of state-of-the-art anti-ship ballistic and cruise missiles within their range of less than 2,000 km, but so far only China has developed such medium-range anti-ship ballistic missiles. U.S. aircraft carriers remain formidable for other countries and at high sea far away from China.

This is a situation similar to that at the end of battleship era when World War II broke out. British battleships that Britain relied on for its world hegemony before the war were no longer able to control the sea within the range of German air force.

The battleships that remained dominant at high sea, were made obsolete by aircraft carriers a few years later.

To Avoid U.S. Blockade of Its Trade Lifelines, China Shall Not Confront U.S.

China has developed enough missile capabilities to defeat U.S. carrier battle groups near its coast. That does not mean that China is able to win in confronting the U.S. U.S. can subdue China without fighting by cutting China's trade lifelines as its navy dominates the oceans. Therefore, quite many Chinese people hold that China needs nuclear aircraft carriers to defend its trade lifelines. That is a false argument. It takes China perhaps three decades to catch up with the U.S. in developing, building and operating nuclear aircraft carrier battle groups equal to those of the U.S. who has experience in building and operating aircraft carriers for more than seven decades.

What China can do now?

Nothing if China confronts the U.S.

China shall maintain good relations with the U.S. and avoid U.S. suspicion that China has any intention to replace the U.S. as world leader; therefore, it shall not build a nuclear aircraft carrier to give rise to the suspect that it has any intention to replace the U.S. as world leader.

In addition, it shall avoid confronting the U.S.

Chinese leaders were wise in removing China's oil rig one month before schedule from disputed waters soon after U.S. Senate unanimously passed a resolution on July 10, 2014 calling on China to withdraw its giant drilling rig and associated maritime forces from disputed waters.

"The enemy advances, we retreat; the enemy camps, we harass; the enemy tires, we attack; the enemy retreats, we pursue." That has long been Chinese communists' strategy in dealing with an enemy much stronger than them.

Not being strong enough to confront the U.S., China retreated.

By so doing, China showed its respect for U.S. world leadership. It was much more desirable for the U.S. than forming a new alliance with Vietnam from which the U.S. will not be much benefited but will have a grave obligation to fight against China when Vietnam wants the U.S. to support it in its war with China.

When the U.S. is in trouble in dealing with crisis elsewhere, China will attack again, i.e. "the enemy tires, we attack." China has not said that its removal of the rig was due to U.S. pressure. A front-page article on Chinese government's mouthpiece People's Daily Overseas Edition claims that China's removal of the rig is a mere coincidence with U.S. Senate's resolution. The writer of article Su Xiaohui gives his article the title "Oil Rig No. 981 Obeys Only China's Order" to refute the view that "China bowed to U.S. pressure" in removing the rig.

I believe that China in fact removed the rig one month before schedule due to U.S. pressure and it is wise to avoid confrontation with the U.S. Sun Tzu teaches us in his *The Art of War* that before going into a war, the commander-in-chief has first to make calculation how many factors there are for him to win the war. If there are not enough factors, he shall not go into the war.

China does not have enough factors to win in its confrontation with the U.S. now. It certainly shall avoid confrontation with the U.S. if its interests are not seriously affected.

In fact the oil rig move is not most important for China in the South China Sea as China can first exploit oil and gas resources in areas without dispute. It can exploit the resources in disputed waters later when Chinese military has grown much stronger to have absolute superiority over U.S. military.

The oil and gas resources are in a vast sea area to be exploited for decades to come. There is no need of haste in exploiting them. What matters is the future. The U.S. is declining while China is rising. If China can keep the trend while the U.S. cannot reverse the trend, the future belongs to China.

China is strong because it has a wise and strong leader due to the emergence of a new generation of talented intellectuals with moral integrity during the Cultural Revolution. I described that in my book *Tiananmen's Tremendous Achievements Expanded 2nd Edition*. My book also describes China's serious problem of succession. When the wise leader is too old to maintain his leadership two decades later, there may be an incompetent Chinese leader or even a despotic leader like Mao in China.

Therefore, if China fails to reform its political system, it will lose steam two decades later.

It is high time for the U.S. to elect a wise and strong leader to replace its incompetent and weak president now. What the U.S. shall be strictly on guard is its ignorance of China's potential and arrogance about its strength.

China's Expedient Measures to Deal with Aircraft Carriers at High Sea

In our missile era now, to counter aircraft carrier combat fleet at high sea, a country need not build nuclear carriers that are too complicated and expensive to build and maintain. The best expedient way is to build stealth VTOL aircrafts and light aircraft carriers to carry VTOL aircrafts. China's experience in building Type 071 landing platform dock (LPD) with 20,000-ton displacement may be useful as the LPD is of the right size for such light carrier, but it has to be installed with more powerful engine to increase its speed from 22 to more than 30 knots.

Such carriers cost much less and take much less time to build. The major drawback is that such carriers are unable to carry large fixed-wing AEW&Cs. China has to develop smaller AEW&C aircrafts or send large AEW&C aircrafts to escort the carriers by means of refuel.

Assisted by enough conventional and attack nuclear submarines, a light aircraft carrier battle group with 3 to 4 light carriers and their supporting warships will be able to deal with a U.S. carrier battle group though without superiority.

China Has to Develop the Strategy of the Space Era

Chinese strategists this time are able to follow the trend. They are wise enough to see that Chinese navy is much inferior to the U.S. and its air force has still to catch up. If the war takes place near Chinese coast, China will not lose as it has a great number of land-based missiles as U.S. Air-Sea Battle cannot deal with repeated saturate attack of Chinese missiles from land, sea and air.

. What worries Chinese strategists is the naval battle far away from China to prevent the U.S. from cutting China's trade lifelines. They have to be able to develop capabilities to deal with U.S. aircraft carrier battle groups far away from China. As they are aware of the world's entry into the space era, they developed the strategy of integrated space and air capabilities for both attack and defense for Chinese air force.

China's top military authority accepted their strategy a decade ago. U.S. outdated Air-Sea Battle strategy certainly has not made them change their strategy. On the contrary, it has enabled them to take advantage of U.S. erroneous strategy to surpass the U.S. in integrated space and air capabilities.

Xi Jinping's Stress on Space-Air Battle Roused Widespread Media Concerns

On April 14, Chinese President Xi Jinping paid a special visit to the offices of Chinese air force and called on the officers there to speed up the development of a powerful air force with integrated space-air capabilities for both attack and defense.

Foreign media vied with one another in reporting his words that "China will strengthen space defense with new combat force" and "China will build up an air force with integrated space and air capabilities".

However, U.S. government and military do not seem much impressed perhaps because China lags far behind the U.S. in its space technology. The U.S. remains satisfied with the strategy it developed in 2008 of Air-Sea Battle to deal with China's military rise. Some analysts believe that China will lose such a battle due to its much inferior navy.

Failure to Notice Two Tremendous Changes in China's Goals

First, China has always stressed its desire for peaceful rise. Second, it has always said that the development of its military is for defense instead of attack.

This time, foreign media seems to regard Xi's idea on strengthening integrated space-air capability as attack rather than defense in nature.

To make clear what Xi really meant in his speech, I gave an accurate interpretation of Xi's words in the following passage: "he called on them (Chinese air force senior officers) to speed up the development of a strong air force with integrated space and air capabilities for both attack and defense."

China Established the Strategy of Space-Air Battle in 2004

In fact, that is not something new. The *Complete Encyclopedia of Chinese Air Force* says, "In 2004, in accordance with the Central Military Commission's guideline on military strategy for the new era, the air force established its strategy of 'integrated space-air capabilities for both attack and defense' and further made clear the core contents of the strategy for Chinese air force...." According to that book, obviously, the strategy is for both attack and defense.

That may rouse widespread concerns in the world. Chinese government's mouthpiece Global Times' report on Xi's visit to the air force tries to ease such concerns. It quotes an anonymous Chinese military expert as saying: Development of integrated space and air capabilities means but enhancing the utilization by the air force of space-based information system to strengthen defense.

Do you believe that?

China's Integrated Space-Air Capabilities Are Not for Defense Only

It is a common military theory that attack is the best defense. It is regarded as positive defense while defense without attack is regarded as passive defense inferior to positive defense.

It is quite natural that a country has to develop attack capability for the sole purpose of defense. However, no one can guarantee that the weapon system a country has developed will be used for defense only. The expert had better not regard people as so naïve as to believe his allegation.

On April 16, China's major official media Global Times said a Chinese military expert unwilling to disclose his identity told its reporter that according to international theory on integrated space-air battle, "integration of space and air" at least includes, but not limited to, first of all the integrated space and air battlefield. There are some physical differences between the spaces within and outside the atmosphere of the Earth, but no absolute border between them. The space and air constitute an integrated whole without any seam of connection. They constitute an integrated battlefield.

Second, the integrated space-air forces: Air and aerospace forces are roughly the same in their organizational structure, training, command and control. Their formations consist of the troops using aviation vehicles, ICBMs, artificial satellites and aerospaceplanes, the troops that intercept aircrafts, ballistic missiles and various missiles from aircrafts and the troops with kinetic weapons and laser weapons.

The aerospace force is in charge of aerospace reconnaissance to conduct global reconnaissance and surveillance. It is also in charge of launching aerospace vehicles to the space and seizing dominance of the space. The air force is in charge of fast transport of troops, carrying out attack in depth and obtaining dominance of the sky.

Finally integrated space-air combat operations are the integration of aerospace and air operations with those of the force and equipment deployed on the ground but directly serve aerospace and air forces.

The expert believes that since the United States put forth its "Asia-Pacific rebalance" strategy and "Air-Sea Battle", China's development of navy and air force has become foreign media's focus of interest. That is the reason why Western media linked China's upgrading of its space defense with China's anti-satellite

(ASAT) tests. He regards it as an instance of sensationalizing China's military development.

China has always claimed that its ambitious space program is peaceful in nature, but such allegation has been questioned since 2007 when Chinese military destroyed a satellite in orbit with a land-based missile.

There may be the explanation that attacking enemy satellite can make it unable to provide guidance for enemy's attack. It is a common military theory that attack is the best defense. As mentioned above, it is regarded as positive defense and it is quite natural that a country has to develop attack capability for the sole purpose of defense. However, Xi's words on development of a strong air force with integrated space and air capabilities for both attack and defense have especially roused widespread concerns in the world. Xi proves that China indeed wants to develop the capabilities first for attack and then for defense.

The expert tried to ease such concerns. He said: Development of integrated space and air strength means but enhancing the utilization by the air force of space-based information system to strengthen defense. As there may be attack of integrated space and air weapon, especially the hypersonic weapons some countries have been developing, China needs to obtain early warning, especially strategic early warning intelligence through its space based information system.

The U.S. Developed Space Strategy Much Earlier

In fact the idea of integrating space and air emerged in mid 20th Century when Russia and the United States contended in the outer space.

In 1959, for the first time the term air capability was replaced by aerospace capability in U.S. Air Force (USAF) Doctrine. The entire space above Earth surface was regarded as aerospace, the battlefield of the air force.

Since then, the term "aerospace" has almost always been used in USAF's various theoretical documents on basic principles. In addition, in a symposium of the U.S. Aerospace Education Foundation in early 1999, American defense experts put forth the new theory that regards joint aerospace force as the mainstay and pointed out that joint aerospace force is a combat force consisting of the aerospace troops from various services and the force shall achieve its combat goal through unified and coordinated operations.

China began discussing the idea of and developing the strategy for integrated space-air battle quite late. As mentioned above Chinese air force established the strategy in 2004.

Will China Be Leader in Space and Leave the U.S. to Be Leader on Earth?

China's development of space-air capabilities and lunar program seems to indicate that.

In missile era, what we need are high speed and accuracy. However, we shall not forget we are now in the space era, which demands much higher speed and accuracy.

Therefore, China is switching a substantial part of its financial resources to its space program to explore first the moon and then Mars. Exploration of Mars requires high speed 10 or 20 times the speed of 11,000 meters per second achieved now. It will reduce the travel to Mars from 7 months to 10-20 days.

Certainly, there are lots of very difficult problems to tackle but it is what the human race has to achieve. The most difficult problem is to achieve high speed and the high accuracy to control high-speed spacecraft to make it slow down and land on Mars and the earth safely.

The development of such technology will enable China to have hypersonic weapons with absolute superiority over aircraft carrier battle groups. The picture on the cover of this book describes how an aerospace bomber destroys an aircraft carrier battle group with hypersonic missiles. It is possible that with abundant fund China may obtain the technology within 2 decades, much earlier than the development of a nuclear aircraft carrier fleet rival to the U.S. Why? Because U.S. success in developing space shuttles proves the technology is already there.

China may even develop flying vehicles quicker than nuclear ICBMs so as to intercept ICBMs in midcourse and forever remove the threat of nuclear holocaust to the human race.

Note: The difficulty of ICBM interception now lies in hitting a high-speed missile with another high-speed missile because the difference in speeds between the two missiles is very large.

There is no problem if the two are close to each other and fly at the same speed in the same direction. That is the case when one person tries to intercept another person close to him when they are running with the same speed in the same direction in an area near equator. Both people are moving along with the earth at the high speed of 1,000 mile per hour, but there is no difficulty for one person with a gun to shoot the other. Their absolute speed is huge but their relative speed, i.e. the difference between their speeds is very little.

An aerospace aircraft with higher speed than that of an ICBM can catch up with an ICBM. Then it reduces its speed to that equal to that of the ICBM when it is close to the ICBM. As they are close to each other and the difference between their speeds is near zero, the aircraft can shoot down the ICBM very easily.

The U.S. may be uneasy if China has switched its focus to the space, but China's space program has not made and will not make the U.S. so nervous as China's launch of a nuclear aircraft carrier. The U.S. perhaps regards space technology as too difficult that China can hardly achieve much progress. However, it has to be aware that unlike the time when the U.S. explored the moon in late 1960s, science and technology are much more advanced now to enable China to make some breakthroughs in space technology.

Knowing that the future lies on how much a country has achieved in space, the U.S. shall make comparable efforts in space so that China will not be the country that dominates the space in the future.

2. Why Will the U.S. Be the Sure Loser?

A2/AD Is a Strategy U.S. Assumed for China

> Know yourself and know your enemy, you will never be
> in peril in battles; know yourself but not your enemy, you have a
> fifty-fifty chance to win; know neither yourself nor your enemy,
> you are in peril in each and every battle.

The Art of War by Sun Tzu

The U.S. knows that well; therefore, it attaches great importance to and makes huge investment in its intelligence capabilities to gather information about potential adversaries.

The U.S. assumes that China adopts what it regards as anti-access and area denial (A2/AD) strategy. That is why China develops anti-ship missiles to deprive U.S. access to and deny U.S. entry to the sea areas around China so that the U.S. applies the Air-Sea Battle strategy to counter China's efforts.

Has any Chinese military leader or strategic document mentioned that China would adopt A2/AD strategy in future military conflict? No. Has U.S. intelligence learnt that Chinese strategy from any secret document of China's Central Military Commissions? No, U.S. intelligence has never been able to infiltrate China's power center. China's strategy to counter the U.S. is its top secret.

Even though we now know that China wants its air force to develop integrated space and air capabilities for both attack and defense, we know nothing about the details of the strategy: What are the weapons to be used? What system of organization to be established? Who will be the commander-in-chief and which unit will be in charge of coordination among various forces?

"The enemy advances, we retreat; the enemy camps, we harass; the enemy tires, we attack; the enemy retreats, we pursue." As mentioned above, that has long been Chinese communists' strategy in dealing with an enemy much stronger than them.

It is not aimed at refusing enemy access or denying enemy entry into the area controlled by Chinese communists. On the contrary, Chinese communists retreat to let the enemy take the area. What is their goal then? Their goal is to annihilate a part of the enemy.

When Chinese communists are strong, their goal remains annihilation of their enemy. A relevant example is the late stage of Chinese civil war when the PLA had grown stronger than its enemy. Their strategy was to annihilate the enemy in northern China so that enemy force would not retreat to the south of the Yangtze River, rely on the river to resist the communists and make preparations for counterattack.

U.S. Has Had Little Success in Countering A2/AD with Air-Sea Battles

It is really a mystery why U.S. strategists fail to learn lessons form their failures in dealing with the A2/AD efforts of their enemies in Vietnam, Iraq and Afghanistan where the U.S. has full air and sea supremacy to fight Air-Sea Battle but without success. Why do they believe they will succeed in dealing with the A2/Ad of China, a much stronger adversary?

A war with China will be entirely different from U.S. previous wars in Vietnam, Iraq or Afghanistan where U.S. troops were trying hard to counter their enemies' A2/AD efforts without success. U.S. troops had to withdraw and equip and train local troops to fight on and then lose the war. It was so in Vietnam. The model of failure was then repeated in Iraq and is being repeated in Afghanistan now.

U.S. disadvantages are obvious. It will be fighting a war far from its homeland without any land support and with a very long distance for transport of supplies and supporting equipment and troops.

You may think that the U.S. can use its facilities in Japan and Guam. No, all those facilities are within the range of China's medium-range ballistic missiles. China has more than 1,000 medium-range ballistic missiles while U.S. technology for interception of ballistic missiles is still far from mature. It has not yet commissioned its interception missiles. Even if it has, it can at best intercept less than a half of the medium-range ballistic missiles. 500 medium-range missiles with high explosive are more than enough to destroy all U.S. and Japanese military facilities.

U.S. strategists fail to realize the simple fact that the U.S. is strong at high sea where the Chinese have no support from their ground-based missiles, air force, radar, etc. The U.S. may subdue China without fighting by cutting China's trade lifelines. Without a navy with aircraft carriers of equal strength, China's effective defense against formidable U.S. aircraft carrier battle groups is but its attack nuclear and conventional submarines. In this regard, China has no advantages at all as U.S. has the strongest submarine fleet in the world and its aircrafts and warships have quite strong anti-submarine capabilities.

The balance of strength is entirely different in the area near China, the U.S. can only rely on its F-22s to dominate the sky but its warships will be annihilated by China's excessively large number of anti-ship ballistic, cruise and other missiles.

However, even F-22s may be in trouble as China has more than 1,000 medium-range ballistic missiles to destroy all the U.S. airfields within the range of the missiles. When China has commissioned sufficient number of its J-20s to counter U.S. F-22s, with the support of land-based radar and EAW&C aircrafts, J-20s can easily grab dominance of sky from F-22s.

From that we can see how stupid Obama's pivot to Asia is. If Obama maintains 50% of U.S. military to counter China, 50% of its military will remain intact to cut China's trade lifelines when the 50% near China is annihilated by China. Transfer of 10% more its military to the area near China will enable China to destroy 60% of U.S. military, leaving only 40% to deal with China at high sea.

However, U.S. misfortune will not be restricted to that. If when the war breaks out, China has successful developed and commissioned aerospace bombers, the remaining 40% will be annihilated by China's aerospace bombers within hours.

The aerospace bomber described in the beginning of this book is not science fiction but very probable reality. U.S. space shuttle is precisely an aerospace plane. If the U.S. had kept on developing it, it would have been able to have such aerospace bombers long ago.

If the U.S. switches to focusing on developing space capabilities now, in the competition for space and air superiority, the U.S. will not necessarily be a loser. After all, it has much more experience than China in developing equipment for space programs, especially the control and soft landing of a space shuttle, which is essential for an aerospace plane.

However, if Chinese economy grows much bigger than the U.S. the U.S. will not be able to maintain its superiority in the long run.

The Mystery of China Surpassing the U.S. Militarily Soon

It is now obvious that Chinese economy will surpass the U.S. Its GDP will surpass the U.S. within a decade if it keeps its current growth rate at a little more than 7%. However, as Chinese currency the Renminbi is now undervalued, if calculated in terms of purchasing power, Chinese economy will surpass the U.S. by the end of 2014.

However, the U.S. still rests at ease that it takes a long time for China to catch up with the U.S. militarily as according to its estimate, China lags behind the U.S. by at least two decades. Really?

In Chapter 6 "China's Arms Race with the U.S.", I list the following four factors that enable China to surpass the U.S. militarily soon:

First, China has a leader with knowledge about not only military strategy but also the technology required in modernizing Chinese military.

Second, Chinese military has an unlimited budget while China has sufficient financial resources to support such unlimited spending.

Third, it can get advanced technology from Europe not only from Ukraine and Russia but also from France, Germany, Britain and other EU members.

The last but not the least, China has talented scientists and engineers who develop weapons for China with patriotic dedication.

There is quite detailed description in that chapter. What I shall elaborate here is the essential factor that enables China to be the strongest in the world.

The most important factor for China to achieve its fast economic growth and modernizations is China's success in giving play to people's talents and diligence.

For example, China lagged the U.S. by more than 20 years in the development of UAV (unmanned air vehicles) also called drone but it has now surpassed the U.S. in certain drone technologies and is expected to entirely surpass the U.S. soon.

China Lagged behind U.S. in Development of Drone but Is Surpassing U.S. Now

The U.S. conducted maiden flight of its UAV Firebee in 1951 and soon began mass production and use of the UAV for reconnaissance.

China shot down a Firebee and began to copy it in 1972 but could not begin using it until 1978 due to technical problems. We can safely say that China lagged behind the U.S. by at least 20 years.

What is the situation now?

China is producing various kinds of drones and exporting some of them.

The U.S. has its well-known reconnaissance drone Global Hawk. China has Soar Dragon with comparable functions and performance. That is not much as Global Hawk is not the most advanced drone U.S. has.

U.S. most advanced X-47B stealth attack drone conducted its maiden flight on February 4, 2011 while on November 21, 2013, China's Lijian Stealth Unmanned Attack Aircraft conducted its maiden flight. China had been catching up quickly and lagged behind the U.S. by less than 3 years in November 2013.

That is not all. China has surpassed the U.S. in certain top new drone technologies.

Qianzhan.com said that on July 13, 2014, U.S. *Defense News* published Wendell Minnick's article "China Develops Mature, Broad-Based UAV Sector". The article describes China's shocking breakthroughs in drone technology.

It is the information leant by renowned UAV specialist Robert Michelson at China's 2011 and 2013 UAV Grand Prix. Michelson, principal research engineer emeritus at Georgia Tech Research Institute, is one of the rare experts who has served as an "International Referee" and "Innovation Forum" keynote speaker at the Prix.

In 2011, Michelson saw demonstration of a "stopped-rotor" vehicle by Northwestern Polytechnic University, Xian, China.

In the 1980s, both the U.S. Defense Advanced Research Projects Agency (DARPA) and NASA funded the Sikorsky X-Wing project for a similar vehicle but failed after significant expenditures.

Boeing spent several years in developing similar technology but also failed. However, in 2011, Michelson saw the impossible in China. The Northwestern Polytechnic University's stopped-rotor UAV "performed flawlessly, transitioning from hover to high-speed forward flight and back again on several occasions."

On Michelson's second trip to China in 2013, he was shocked by Nanjing University of Aeronautics and Astronautics' creation of "PlasMav" that used a high-voltage field to reduce drag and increase lift. It is something never attempted on a UAV before.

In the U.S. development of new technology and equipment is mostly done by big established enterprises.

China, however, knows talented scientists and engineers such as Steve Jobs and Bill Gates displayed their gifts when they were very young. The prix is a way to discover talents. Certainly, there are other ways to find talents in China. The appointment of a young girl Yang Yi as chief designer of China's new fast attack boats is another example. The boat is regarded by U.S. military experts as aircraft carrier killer. The prix and Type 022 fast boats will respectively described in details in Chapters 13 and 9.

3. Obama's Poor Stratagem and Diplomacy

Obama Unaware of U.S. Losing World Leadership

Amid much criticism at home against his poor leadership and diplomacy, Obama made a speech at West Point on May 28, 2014, claiming that he has not cause the U.S. to lose world leadership but has enabled the U.S. to maintain world leadership for 100 years to come.

With his brand eloquence, he seemed quite convincing. His top argument was that the U.S. remains the strongest economically and militarily. He is ignorant that one leads by wisdom instead of strength and war is not won merely by strength.

One can be regarded as the leader of a group if the group of people follows his leadership. No matter how strong he is in the group, if most of the people in the group follow another man's instead of his leadership, the another man is the leader of the group. If no one in the group follow any one's leadership, however strong one is in the group, the group remains leaderless.

Requirements for the Leader: High Moral Standards and Awe

As a leader, one has to be on moral high ground so that all the good guys will follow him. One has to create awe among the bad guys so that they dare not create trouble in the group. There will thus be good order in the group and all the guys that are not bad will be happy and have confidence in and follow one's leadership.

Prevention of bad guys from creating trouble so as to maintain good order is much more important. Only by so doing, can one make necessary contributions to the group. Members of the group will follow one's leadership as his leadership is indispensable for their peace and security.

Since America's defeat in Vietnam, People had lost confidence in U.S. military power and dared to disregard it. However, the U.S. was lucky as its rival the Soviet Union soon sought improvement of its relations with the West and then disintegrated.

Later, the U.S. shocked the world in its swift victory in the first Gulf War and created the awe of U.S. invincible power throughout the world. If U.S. leaders had been aware of the importance in keeping the awe to maintain its status as the world police other countries had to respect, it would not have been in trouble later.

Winning each and every battle is not the best of the best;

subduing the enemy without fighting is the best of the best.

The Art of War by Sun Tzu

Having created the awe, the U.S. can easily subdue its enemy by the mere threat of war. It can follow Sun Tzu's teaching to subdue the enemy with the threat of war without actually fighting a war.

However, knowing the importance to maintain the awe, the U.S. has to be very careful that whenever it fights a war, it must be sure to win the war and achieve its goal quickly. By so doing, it will enhance the awe and make its threat of war more effective in preventing future war so that it will be able to subdue the enemy without fighting.

The U.S. precisely acted that way in the first Gulf War. As soon as the U.S. achieved its goal to punish the aggressor, it retreated to maintain regional balance.

I mentioned in my book *Tiananmen's Tremendous Achievements Expanded 2nd Edition* that the awe created by the Gulf War was so shocking that it helped Jiang Zemin establish his authority over China's People's Liberation Army.

U.S. leaders must have the basic knowledge that however they love wars, U.S. people are peace loving and hate wars. Therefore, they shall refrain from fighting any war but maintain the awe already created so that no other countries dare to create trouble.

It is a pity that U.S. leaders were carried away by U.S. victory in the Gulf War and became so arrogant as to fight two wars simultaneously. The failure in the two wars is disastrous. The U.S. is heavily in debt and has lost the awe completely.

People now know that in spite of U.S. powerful military, the U.S. has its Achilles heel. As U.S. servicemen do not like war, the U.S. has to spend lots of money to provide the best equipment for them. As a result, war is too expensive for the U.S.

Due to U.S. failures in Iraq and Afghanistan, U.S. threat of war is now useless. It was countered by Putin's threat of nuclear war during the confrontation between the U.S. and Russia over Ukraine issue. It was challenged by China's establishment of its East China Sea Air Defense Identification Zone. Even the very week Islamic extremists without any weapon industry dared to boast that they would attack the U.S.

What make U.S. predicament worse is that U.S. leaders know nothing about diplomacy. On July 2, 2014, Reuters publishes a report titled "U.S. military to face growing crises, falling budgets" to describe the predicament the U.S. is in two months after Obama boasted that America remains world leader and will remain so for a century to come.

Reuters begins its report by saying, "First it was worries over the South China Sea, Afghanistan, Libya, Mali and Syria. Then it was Russia's annexation of Crimea and the hunt for Nigerian schoolgirls kidnapped by Boko Haram. Now the United States and its allies find themselves preparing once again for potential military action in Iraq."

I am more impressed by the most interesting photo at the beginning of the report, showing how frustrated U.S. servicemen are in their drill with corrupt and incompetent Philippine troops. The photo can be viewed at https://www.linkedin.com/today/post/article/20140703061457-62667533-ll-ll?trk= mp-author-card.

That is okay for the Philippines as it need not strive to be rich and strong. It has a rich and powerful big brother who will help it when it is in trouble though the Philippines is no small country with a population one third of the size of the United States.

Obama knows how to talk, but not how to lead

Obama does not know that to be global leader he shall have strong and trustworthy allies.

China used to be his trustworthy ally. The most convincing evidence was China's unselfish support for Obama's actions in Libya. China suffered serious losses for that.

However, Obama looked down on such a trustworthy ally and turned it to Russia's side by his foolish pivot to Asia aimed at containing China.

In fact, the U.S. is unable to contain China now. Whether China will grow strong and remain prosperous depends on its own ability to deal with its own problems. In this respect, the U.S. can neither help nor hinder.

As for China's disputes with its neighbors Japan, the Philippines and Vietnam, the U.S. could not help them defeat China because the area is too far away from the U.S. but right in front of China. China can use its land-based aircrafts and missiles to deal with the limited number of U.S. aircrafts and warships.

In July 2012, the Strategic Studies Institute (SSI) of U.S. Army War College published a long report titled "The Next Arms Race" edited by Henry D. Sokolsky.

Its Chapter 5 "China and the Emerging Strategic Competition in Aerospace

Power" by Mark Stokes and Ian Easton, responsible persons of the 2049 Project Research Institute, describes the threat of China's development of advanced ballistic and cruise missiles, especially the anti-ship ones to U.S. navy. It concludes by mentioning the shift in U.S. basin foreseen by Robert Kaplan of the Center for a New American Security — moving away from allied territories to Guam and the South Pacific Islands—and a greater U.S. naval presence in the Indian Ocean.

If one fails to follow the development of technology, one may not believe that. The era of gunboats has long passed. Since the 1950s, it was the era of missiles. Missiles are good to attack but difficult to defend. In spite of U.S. huge efforts, the success rate of intercepting ICBMs has not been satisfactory so far.

It is cheap to produce lots of cruise missiles to attack the huge and slow target of an aircraft carrier, but it is difficult and expensive to intercept small missiles at high speed. The odds are obviously on missiles. The U.S. can never reverse that reality of technological development. The only way out is to follow the trend. What is the trend now? It is now the space era. The U.S. has to follow the trend to develop integrated space and air capabilities instead of sticking to its obsolete Air-Sea Battle.

Obama's Poor Diplomacy

In the area of diplomacy, America began to lose world leadership because as mentioned above it not only lost China as its vital ally but also has turned China to Russia's side to form a Russian-Chinese alliance to counter the U.S.

While U.S. talented statesmen and diplomats want both U.S.-Chinese and U.S.-Russian relations to be better than Russian-Chinese relations, Obama is pursuing the contrary.

Obama is correct to be concerned about China's rise as China is not a democracy and was ruled by a despot Mao Zedong not very long ago. However, in U.S. relations with China, like its relations with quite a few other countries, there are things the U.S. shall exploit for its benefits and things that may harm U.S. interests that it shall prevent.

China is very useful for the U.S. in containing Russia and North Korea. In terms of diplomacy, most harmful and even dangerous to the U.S. is Chinese alliance with Russia.

China's rise is what the U.S. shall be concerned about, but the U.S. can do nothing to prevent China's rise. It can only rally other world powers to contain

26

China.

In this regard, Obama fails to see that not only his allies Japan and South Korea but also Russia and India are concerned about China's rise. He would have had good opportunity to rally Japan, Russia, India and South Korea around the United States in dealing with China if he had had the skill to be world leader.

Among the four potential allies, Russia, though declining, is the strongest militarily. It still has the technology very useful for China's military modernization but due to Russia's concern about China's rise, Russia refused to transfer the most advanced military technology China wanted. It only wanted to sell China advanced weapons to get funds for its own military ambition.

Obviously, Russia did not want to live under the threat of a militarily powerful neighbor; therefore, U.S. containment of China will benefit Russia.

It is a common mindset that a previously powerful country or family has an earnest desire to restore its past glory. Putin knows that well. He knows his autocracy is unpopular but by his efforts to restore the past glory of the collapsed Soviet Union, he will make himself popular.

That is why he has drawn up an ambitious plan to establish a powerful navy to restore Russia's position as a superpower.

To prevent Russia's restoration of its position as a rival to the U.S., the U.S. needs China. In fact, like China's rise to Russia, Russia's rise may constitute a real threat to China.

In dealing with Russia, China has the traditional wisdom of allying with remote states to attack neighboring states. It was Fan Sui's well-known strategy adopted by the State of Qin to gradually conquer all other states and unify China more than 2,000 years ago.

The strategy is so well-known in China that when Mao decided to improve relations with the U.S. in early 1970s, nearly everybody I knew who had had some knowledge about Chinese history commented that Mao adopted Fan Sui's strategy of "allying with remote states to attack neighboring states".

In Zhisui Li's memoir about Mao, he says Mao told him Mao's move was based on Fan Sui's strategy of "allying with remote states to attack neighboring states".

It is very clear Chinese leaders adopted that traditional strategy to ally with the U.S. to contain Russia when they supported the West's military actions in Libya that would weaken Russian influence in the Middle East.

With such background of Russia's concern about China's rise and China's traditional strategy, an alliance between Russia and China was highly impossible. However, Obama has made it possible.

When Obama had announced his pivot to Asia and begun to make things difficult in China's maritime territorial disputes with its neighbors, China began to regard Russia as the only possible ally with enough strength to counter the U.S.

China filled Russian President Putin with joy when it joined Russia in vetoing the UN resolution on Syria proposed by the West

The two countries began to establish a Cold War alliance. To show the importance attached to mutual relations, Putin and Chinese President Xi Jinping each chose the other's country as the first destination of their visit abroad after they were elected. However, due to lack of mutual trust, quite a few deals made during their visits have not been completed until Putin's visit to Beijing on May 20, 2014. For example, Russian sales of Su-35 fighter jets, Lada class submarines and S-400 air defense missiles, and Sino-Russian cooperation in making large airliners and heavy helicopters.

Close cooperation between China and Russia in weapon development will have the consequence of adding wings to tigers for both countries. That will be what the U.S. fears most.

But Obama did not fear, the U.S. squandered lots of opportunities to break the fragile Sino-Russian alliance at its burgeoning stage.

Russia has sent its experts to Shanghai for joint development of aircrafts. The project, if successful, will grab a large market share of airliner market from Boeing and Airbus. However, there will be lots of problems related to sharing of intellectual property and technology, pricing, sales and distribution of profits. No one can be sure the project and other major projects between Russia and China will succeed. However, Obama came out to help Russia and China.

In Obama's interview with Economist magazine in early August 2014, he said, "I do think it's important to keep perspective. Russia doesn't make anything,"

For Russian people who want to restore their country's past glory, Obama's contempt is the greatest insult. Nothing Obama has ever said may better enhance Russia's desire to ally with China.

Russia Doesn't Make Anything Alone but Makes Lots Allying with China

True, Russia really doesn't make anything alone. It's too weak to counter the U.S. So is China though China's economy is much bigger than Russia's.

Obama can look down upon Russia and China when they are separated but cannot look down upon any of them when there is an alliance between them.

Jointly, they are strong enough to counter the U.S. For example, Russia makes Su-35 fighter jet, S-400 air defense missiles, Lada-class submarines, etc. can help China shoot down U.S. aircrafts and sink U.S. warships if there is a war between U.S. and China.

If the U.S. decides to use nuclear weapons, how can it be sure that after its first strike at China, Russia will not conduct first strike at the U.S. followed by Chinese second strike.

Regarding China, Obama told the Economist that the U.S. had to be pretty firm with China to make China "meet resistance" and stop pushing hard. He is pushing China hard to Russia's side.

China can provide Russia with funds, lots of supplies and also advanced anti-satellite, hypersonic and drone technologies when there is a war between Russia and the U.S.

Sino-Russian alliance is strong enough to defeat the U.S. if the U.S. has no allies to fight by its side.

I do not know why Obama almost devoted entire interview in strengthening Russian-Chinese alliance. For example, he said, "Immigrants aren't rushing to Moscow in search of opportunity.... The population is shrinking."

Chinese people are rushing into Siberia in search of opportunity but Russia restricts their entries. What if in response to Obama's insult, Russia allows Chinese people to become Russian citizens in Siberia if Chinese people help Russia exploit the rich resources there. Russia can precisely provide an outlet for China's surplus population to make up for its shrinking population.

How can the leader of a superpower be such an illiterate in diplomacy?

Does Obama Want to Be a Lone Swordsman fighting All Evils in the World Alone?

Japan, South Korea, Russia and India are sufficient to counter China's rise even without U.S. pivot to Asia if only the U.S. can rally them together, but Obama seems to want to deal with China alone. Oh, not alone but burdened with untrustworthy Japan and unfaithful Philippines.

Besides pushing Russia and China together to form an alliance, the U.S. failed to prevent Japanese Prime Minister Shinzo Abe's visit to Yasukuni Shrine that has caused South Korea to sever its relations with Japan. This has provided China opportunity to win over South Korea to its side.

China and India have been enemies since the war between them in 1960s. It provides the U.S. good opportunity to win over India to its side. However, India is close to Russia. Russia is trying hard to form a Russia-India-China alliance.

To put an end to enmity, China has been trying hard to improve relations with India while Russia is helping it doing so. The U.S., however, failed to bet on both sides in India. It snubbed Indian opposition leader Modi for a long time. Now, it is very difficult for the U.S. to improve its relations with Modi when Modi has become Indian Prime Minister.

China, however, received Modi when he was not in power as if he had been state leader during Modi's four visits to China. This will facilitate resolution of the border disputes between the two countries.

If the disputes have been resolved, Russia-India-China alliance may become a reality to greatly reduce U.S. influence in Asia.

The U.S. cherishes Japan as its major ally in the world, but Japan is by no means a trustworthy ally for the U.S. Vice President Biden spent one hour but failed to persuade Japanese Prime Minister Shinzo Abe from visiting Yasukuni Shrine. The visit has created South Korea's bitter enmity against Japan. Japan has thus foiled U.S. plan to form an alliance with Japan and South Korea to counter China.

One is not the leader because of his talks about his leadership nor because he is the strongest militarily or economically. One is the leader because others follow his leadership.

Three powers China, Russia and Japan have refused to follow U.S. leadership on major issues of Ukraine, South China Sea and Yasukuni visit. Can Obama keep on boasting U.S. global leadership?

The situation has grown from bad to worse. With China as a strong ally, Russia has been trying hard to regain the leadership of the former Soviet Union first by openly challenging the U.S. in protecting Snowden and then by taking away Crimea from Ukraine.

Russia has recently set up its new Eurasian Economic Union, which together with the Shanghai Cooperation Organization and the Conference on Interaction and Confidence-Building Measures in Asia (CICA) forms a growing little clique under Russian leadership.

Obama talked much how with sanctions he has contained Russia. However, have his sanctions made Russia return Crimea to Ukraine? Can he be sure that other parts of Ukraine will not split from Ukraine due to his sanctions?

Sanction is indeed a powerful weapon, but one needs trustworthy allies to cooperate with him in imposing his sanction.

The world needs U.S. leadership, but Obama lacks the qualities to be world leader. Domestically, he has not been able to overcome U.S. economic problems to give U.S. the economic strength to be world leader. Internationally, he does not know to keep trustworthy allies but attach importance to mean allies that remain close to the U.S. when they need the U.S. to fight for their interests but drive the U.S. away when they think that the U.S. is no longer useful.

Why is the U.S. in trouble? Because it prefers poor allies and rejects rich or strong allies. Its allies, though not big, are heavy burdens on it.

Why the U.S. cannot have strong and powerful allies?

Because its leaders wants to be world leader for 100 years so that they are jealous whenever other countries become strong.

The jealousy is so strong that the U.S. even tapped German Chancellor Angela Merkel's mobile phone.

Before the U.S. Turned China to Russia's Side, China was U.S. Useful Ally

Russia is the only country who wants to become a rival to the U.S. while China wanted to be U.S. ally before the U.S. turned China to Russia's side by its pivot to Asia though at that time the U.S. did not regard China as its ally.

I would like to regard it as China's one-sided alliance with the U.S. It sounds wield, but it is true.

When the West wanted to have a UN Security Council Resolution on attacking Libya, China fully supported in spite of the heavy losses it might suffer due to China's substantial interests in Libya. It could not help withdrawing more than 50,000 of its citizens from Libya. The mere size of the withdrawal compared with Libya's small population proved how much interest China had there.

West's military actions harm Russia's interest there, but due to Chinese support for the resolution, Russia did not even dare to veto it!

It proved how useful China is for the U.S. when it regarded itself as U.S. ally.

Once China was driven to Russia's side, the U.S. lost its world leadership. It could not get what it wanted related to Syria at the UN Security Council due to joint Chinese and Russian veto. However, Chinese support for Russian veto was only a gift to Russia for the commencement of alliance. China has had no significant interest in Syria to cause it to veto.

Still, Obama was not aware of the trouble Sino-Russian alliance may bring about to the United States. It has made the U.S. world leadership ineffective, but Obama lacks the wisdom to realize that even now when some of the harm has already been done, such as Russia's annexation of Crimea, which Russia would never have had the courage to do if it did not have China as its ally.

More harm will be done in the form of Russia's cutting more parts away from Ukraine.

Now, it is high time for the U.S. to take China as its ally to help the U.S. restore its world leadership. It is common diplomacy that when a country becomes strong to challenge the U.S., the U.S. has to form alliance with other countries, especially the strong ones to counter the challenger. However, when the alliance has succeeded in subduing the old challenger but one of the allies becomes strong and begins to challenge the U.S., the U.S. has to form another alliance perhaps including the old challenger to deal with the new challenger.

That is diplomatic common sense. There is no eternal ally or eternal enemy. Only those who have no historical knowledge at all may lack such common sense.

China Urges U.S. to Accept It as an Ally ahead of Key Meeting

A week before the Sixth Round of China-U.S. Strategic and Economic Dialogues and the Fifth China-U.S. High-level Consultation on People-to-People and Cultural Exchanges scheduled on July 9 and 10, 2014, Chinese President Xi Jinping said: the United States need to "plant more flowers, not thorns" in their relationship and Washington needs to have a more objective view about China.

Xi said that when he met former U.S. Treasury Secretary Henry Paulson and expressed his hope that the two countries would keep "injecting positive energy" into their relationship through such meetings.

There are no substantial conflicts between China and the U.S.

Obama complained about intellectual property in the abovementioned interview, but China is indeed making efforts to protect intellectual property. Only it takes time for China to overcome corruption and official despotism in protecting intellectual property.

China and the United States are closely related in trade and business and cooperated in dealing with important international issues such as North Korea and Iran. However, they differ over quite a few matters from human rights to the value of Chinese currency.

On July 1, 2014, U.S. Treasury Secretary Jack Lew said the value of Chinese currency is a "very big issue" and will demand further appreciation of the value when he attended the Strategic and Economic Dialogue in Beijing. However, China has been making efforts in that respect and allowing its currency to rise in value for more than 30%. Chinese enterprises have to adapt to the rise; therefore, the rise has to be gradual step by step.

U.S. Secretary of State John Kerry will perhaps also attend the dialogue. The hot issues to be discussed will likely be China's territorial disputes with Japan, Vietnam and the Philippines. In fact, no U.S. interests are involved in those disputes. U.S. support for those countries aims merely at containing China.

As it takes a long time for Xi Jinping to conduct his reform and even longer for the reform to take effect, the harm may be caused by China's rise will not emerge in the near future. The U.S. shall break Sino-Russian alliance now and use its alliance with China to impose its world leadership.

When there is any sign that the harm will emerge, it will be able to exploit China's neighbors' concerns about China's rise to rally them including Russia, Japan, India and South Korea to contain China.

4. China Surpasses U.S. in ASAT Capabilities

The U.S. Has Been Monitoring China's Anti-satellite Capability Closely

On January 14, 2013, Reuters published a report titled "China's space activities raising U.S. satellite security concerns". It described U.S. concerns about China's growing ability to disrupt U.S. military and intelligence satellites.

A U.S. source not authorized to speak publicly revealed to Reuters a classified U.S. intelligence assessment in 2012 of China's increasing activities in space.

The intelligence report warned that China was able to disrupt U.S. satellites that provide GPS services, military reconnaissance and communications and early missile warning.

Due to the heightened concern, the U.S. had been watching Chinese activities closely and has been urging China not to repeat its January 2007 test that destroyed a satellite in orbit and created over 10,000 pieces of debris that pose a threat to other spacecrafts.

However, the U.S. did not lag behind China.

The United States had continued to develop its own anti-satellite capabilities. In February 2008, it fired a missile and destroyed an ailing American satellite in orbit.

U.S. Expert Worries about China's ASAT Capability

On April 21, 2014 Michael Austin, a resident scholar at the American Enterprise Institute and a columnist for wsj.com, said in his article titled "China Takes the Fight to Space" on The Wall Street Journal, "The 21st-century battlefield in East Asia might not be on water or air, but in space."

U.S. pivot to Asia means transfer of more military assets to Asia, i.e. 60% instead of 50% of U.S. military strength will be in Asia in the future. The strategy for the pivot is Air-Sea Battle.

Mr. Austin says in his article, "Chinese President Xi Jinping has provided evidence that military competition may be shifting from ships and planes to space. Xi showcased China's military priorities when last week he encouraged his country's air force to better integrate its air and space capabilities." Mr. Austin warned the U.S. to be prepared for China's anti-satellite attack that may cripple U.S. Satellite systems.

He pointed out: China's space program is controlled by its military while in the U.S. military and NASA are separated in organization and leadership. Chinese way of organization has facilitated China gaining technological advantages over the U.S.

He was concerned about China's ability to destroy a satellite in orbit and test of a new, mobile-launched anti-satellite missile, but he would have been even more worried if he had been aware that China had launched three ASAT satellites by one rocket, one of which has a robotic arm that can capture another satellite.

China's manned mission to the Moon also worried him as he was afraid that China might set up a lunar base in the future.

He called on American homeland security and military planners to be concerned about China's growing technological expertise that might make the space a new battlefield. China now has ASAT capability to disrupt U.S. GPS and telecommunications systems with catastrophic effect. "That would impose huge costs on the American economy and potentially shut down entire industries," he said. "It is the 21st century version of saturation bombing, designed to target civilians and break their will to resist."

That will also deprive U.S. military the global advantage of its networked battle system. In addition, U.S. military will be blind without surveillance satellites and GPS system and unable to conduct its Air-Sea Battle against China.

U.S. congress is so concerned about such ASAT attack that Pentagon is requested to study U.S. military's ability to operate when communications is denied.

Moreover, according to Mr. Austin, China will certainly conduct cyber warfare while blinding U.S. satellites. It will effectively paralyze U.S. computer networks and make U.S, forces unable to conduct its operations in Asia and beyond.

When U.S. surveillance satellites have been blinded, U.S. decision makers will be deprived of the access to timely intelligence. As a result, China may take blitz actions to grab disputed islands before the U.S. has time to respond. That may "give Chinese forces the margin of success necessary to present Washington with a fait accompli," Mr. Austin said. "It would undoubtedly be even easier to isolate and interrupt the military activities of America's smaller allies, leaving Washington with a heavier burden."

Mr. Austin's greatest concern is the trend that the reduction of U.S. military budget may cause the U.S. to lose its leadership in space while China is enhancing its integrated space and air capabilities.

The trend is clear: President Obama may want to shrink the military and surrender America's lead in space, but China is moving in the opposite direction and will eventually gain the upper hand. As the U.S. fails to have a clear view of the potential of space capabilities, it will gradually lose its advantages in space in spite of its technological superiority.

China's ASAT Satellites Able to Rendezvous, Capture and Destroy Satellites

I have mentioned that China shot down a satellite in orbit and launched three ASAT satellites, one of which has a robotic arm able to capture a satellite.

Washington Free Beacon gave a detailed report on the three satellites that are especially worrisome for the U.S. They were launched on July 20, 2013 by a Long March-4C rocket. The U.S. found that they conducted some unusual maneuvers in space that gave the impression that they were advanced ASAT satellites.

One of the satellites had an extension arm able to attack satellites in orbit. It is a brand new way to attack a satellite other than kinetic and electronic destruction.

"This is a real concern for U.S. national defense," a U.S. official familiar with intelligence about satellites said. "The three are working in tandem and the one with the arm poses the most concern. This is part of a Chinese 'Star Wars' program."

After some details about the 3 satellites had been disclosed on the Internet by space researchers, the anonymous official revealed some information on China's ASAT program.

He said, "The retractable arm can be used for a number of things – to gouge, knock off course, or grab passing satellites." According to him, the satellites could also repair orbiting satellites.

According to the information on the Internet, the three satellites moved all over the space and could go near other orbiting satellites. Space analysts wondered whether they were China's new space weapons for intercepting, capturing and destroying orbiting satellites.

A space researcher found that on August 26, one of the three satellites lowered its orbit by 93 miles and then moved to rendezvous another satellite at a close distance of merely 100 meters.

Chinese state-run media said that the three satellites were Chuangxin-3 (Innovation-3); the Shiyan-7 (Experiment-7); and Shijian-15 (Practice-15), the one believed to have a robotic arm.

A Pentagon spokesman said that relevant government agency had monitored the satellites as usual but failed to mention the threat to U.S. satellites that the three satellites posed.

The above-mentioned U.S. official believed that the Obama administration was keeping details of China's ASAT capabilities secret in order to play down their threats to U.S. national security. If American general public had learned the threat they would demand an increase in U.S. military budget, which the Obama administration was unable to.

China's lunar space program is a civil program on the surface but the ability to track, identify and rendezvous with satellite obtained in the program are useful in China's ASAT technology as proved by the said three satellites. Obviously, Chinese military is acquiring lots of technologies to improve its space and counter-space capabilities through its lunar program.

In fact, before the launch of the three satellites, China conducted a test of maneuvering small satellite in 2010. Two Chinese satellites rendezvoused several hundred miles above Earth in August 2010.

The Pentagon said at the time, "Our analysts determined there are two Chinese satellites in close proximity of each other. We do not know if they have made physical contact. The Chinese have not contacted us regarding these satellites."

It appeared one of the satellites made contact with another satellite causing it to change orbits. The two satellites were estimated to have been as close as 200 meters to each other.

China's High Direct Ascent ASAT Capable of Reaching Synchronous Orbit

At the end of January 2014, Washington Free Beacon reported that China had recently launched small satellites that are very difficult to track that can interfere with satellites at different altitudes.

It said that the congressional U.S. China Economic and Security Commission said in its latest annual report that on May 13, 2013, China launched a missile that went into high-altitude but did not place a satellite in orbit.

"Although Beijing claims the launch was part of a high-altitude scientific experiment, available data suggest it was intended to test at least the launch vehicle component of a new high-altitude anti-satellite (ASAT) capability," the report said.

China's Electromagnetic Pulse ASAT Capabilities

China Surpasses the U.S. in Development of Electromagnetic Gun

On October 8, 2013, a committee of international experts held that the National High Magnetic Field Center in China's Huazhong University of Science and Technology has already generated one of the best pulsed fields in the world, especially, its design of power source and magnet technology have successfully achieve world first ranking and surpassed by far similar design and technology in the world.

Regarding to this, some military experts hold that the technology is very useful in China's military industry and the electromagnetic gun from that technology will be a high tech weapon under development. Such gun can be used not only in space-based anti-missile system to destroy enemy low-orbit satellites and missiles but also in intercepting the missiles launched from warships and armored vehicles. In addition, it can be used in air-defense systems and anti-armor weapons and in improving conventional guns.

Regarding to that technology, in 2010, the U.S. developed as a powerful weapon its electromagnetic railgun. In a test firing, the railgun could reach Mach 5 and hit a target 200 km away in an instant. It has a range more than ten times of that of a conventional naval weapon. China's success again in that important project indicates that China has surpassed the U.S. in related technology. The abrupt rise of China's military industry scares lots of Western countries.

The space intelligence section of Japan Self-Defense Forces revealed in its newest intelligence: China has destroyed the control chip of a Japanese spy satellite with a secret weapon when the satellite was tracing a Chinese J-20 stealth fighter jet in northwestern China. The satellite is the third Japanese spy satellite launched from Kagoshima, Japan.

U.S. analysts believe that China used electromagnetic pulse weapon Poacher One in the attack. That was China's top secret military research and development project. PLA's electromagnetic weapon Poacher One is able to transmit electromagnetic pulse of several megawatts continuously for one minute to destroy all military and civil electronic information and communications systems operating within a few kilometers. It can also destroy enemy's internal chips.

Other ASAT Weapons and the Potential to Intercept ICBMs

Previously, U.S. military revealed that the PLA sent a satellite near a U.S. spy satellite to blind it with spray of coating on its camera. PLA has lots of means to attack and interfere with satellites. U.S. military is concerned that neutralization of U.S. satellites by PLA's space force will be its nightmare in war.

However, the development of ante-satellite technology does not stop there. It may be the basis for the technology to intercept an ICBM. That will be a much greater worry for U.S. military.

China's Laser ASAT Capabilities

According to U.S. military, there were previous incident of Chinese military blinding a U.S. spy satellite with laser. It was China's move to display its military strength in space.

Huanqiu.com said in its report that according to a report on the website of S. Rajaratnam School of International Studies of Nanyang Technological University, Singapore, under the guidance of its State High Tech Research Development Plan (i.e. Plan 863), China is developing three mystic high technologies for both military and civilian application.

The first is the technology of inertial confinement nuclear fusion laser. Different from common large chemical laser, it uses inertial confinement nuclear fusion to make high-energy laser. This is a new type of small solid laser generator much smaller and more powerful. The United States has applied that technology in developing its high-energy laser anti-ballistic missile and anti-satellite technology.

China has been doing research in this field for years and developed its Shenguang (divine light) 1 and Shenguang 2 large solid powerful laser generation equipment. China Academy of Engineering Physics is now building the infrastructure of Shenguang 3 high-energy research center in Mianyang, Sichuan Province. The project, if completed, will be of great strategic significance in speeding up the development of China's next generation of thermonuclear weapons and promoting China's development of powerful laser weapons.

The second is the technology of hypersonic glide vehicle, which will be described in details later in Chapter 7.

The third is its second generation of Beidou satellite navigation system, which will be described in details at the beginning of Chapter 5.

Later on September 16, 2013, U.S. Defense News website reported that in order to destroy U.S. satellites, China has always been developing anti-satellite laser and missiles.

Michael Raska, a research fellow at the Singapore-based Institute of Defense and Strategic Studies, said that China was carrying out the Shenguang laser project that aims to use high-powered lasers to generate a sustained nuclear fusion reaction.

Raska said that the program, which was officially referred to as an alternative energy program, might be used for two military purposes: improving China's next-generation thermonuclear weapons and speeding up the development of China's directed-energy laser weapon.

According to U.S. military, there was a previous incident of Chinese military blinding a U.S. spy satellite with laser. It was China's move to display its military strength in space.

China a Step ahead of U.S. in ASAT Defense

U.S. Defense Advanced Research Projects Agency (DARPA) was the first to have the idea of quick response satellite system for ASAT defense.

Since China has obtained ASAT capability to neutralize or destroy enemy satellites, it is also well aware that it shall have the ASAT defense capabilities; therefore, it has made great efforts to develop its space quick response capabilities for the situation where its satellites are neutralized or destroyed.

For various reasons, however, the United States has not implemented the idea to set up any completed quick response satellite system except for the launches of some static satellites and small solid-fuel rockets. As the preparations of such a rocket and installation of such a satellite take time at a fixed site, the rocket and satellite may well be destroyed by the enemy before they are launched.

At about noon September 25, 2013, China used a Kuaizhou (quick-vessel) small carrier rocket to launch Satellite Kuaizhou I smoothly into its preset orbit.

It is said that Satellite Kuaizhou I will mainly be used for emergency disaster monitoring and information support for disaster rescue and relief. Its user is the National Remote Sensing Center of China's Ministry of Science and Technology.

It seems a civil project.

However, in future space war, once a Chinese satellite is destroyed by the enemy, it can promptly be replaced through the quick response satellite launch system and turn the adverse situation into a favorable one in the battlefield. Analysts say that Kuaizhou small carrier rocket is a part of the new generation of space quick response battle system that China is developing. At present, only two countries China and the U.S. are developing such systems.

Space quick response battle system is mainly used in the quick launch of satellites or anti-satellite weapons. The successful launch this time marked China's victory in its competition with the U.S. to become the first country that has tested a complete quick response space vessel with integration of a satellite with a rocket. It is of great strategic significance.

In 2002, in order to maintain its world leadership and control the space, the

U.S. began to develop its space quick response system so as to turn the above-mentioned idea into reality. In 2013, China's official media also began to disclose times and again similar battle system. Certainly, other countries lack the technological and financial resources to do so. No wonder, some media believe that the competition between China and the U.S. for monopoly of space superiority will thoroughly change world military situation because in the 21st century, the one who controls the space will have the initiative.

The idea of the Kuaizhou vehicles is to provide emergency services for tactic reconnaissance and communications regarding to specific targets. When a launch platform hidden in a tunnel has received an order, the integrated rocket and satellite movable on a truck will leave the tunnel to prescribed launch site, complete the preparations for launching and launch the rocket quickly to send the satellite to preset orbit to provide required services. It takes a few days or merely hours after receiving the order to put the satellite in service. Conventional launching of such a satellite usually takes 6 to 9 months.

There is the speculation that the size of a Kuaizhou integrated rocket and satellite ranges between a DF-21 and DF-31 missiles. When it leaves the manufacturer, the integrated satellite, rocket, launching tube and truck look like a missile. The rocket and satellite stored in the sealed launching tube may remain effective for about a decade. As its cost is low, China may produce quite a lot of them and keep them in its tunnels. When there is a war, it can launch as many of them for emergency needs like its medium-range missiles.

The satellites in GPS and telecommunications systems are operating in high orbits. Those satellites are expensive and complicated to launch, but cover large areas on earth. Even a spy satellite has to operate in relatively high orbit so that not too large number of them can cover the entire areas they monitor. When they have been destroyed in war, it is impossible to replace them within a short period of time, but their services are indispensable on the battlefield.

To satisfy the need for reconnaissance, positioning, precision guidance and communications, there have to be relatively cheap substitutes to provide short-term services at the battlefield. Those substitutes will operate at very low orbits. As they are not static, there have to be quite a few of them in similar orbit to keep a relatively constant distance between the neighboring two so that they can relay the information to the one flying over the battlefield to provide the services of the high-orbit ones. As the battlefield they cover will not be very big, low-attitude satellites will do. As there is no need for each of them to have many functions, they may be quite small to be launched by mobile rockets. It certainly takes time to develop such a system of substitute satellite system to provide the services for the battlefield. China has been working on it while the U.S. has remained idle.

Integrated Space and Air Capabilities Are Not Merely ASAT Capabilities

If China's new strategy is but the development of ASAT capability, there is indeed nothing new. The U.S. has long been developing such capability. What worries it is that there is now a rival.

Mr. Austin fails to know that integrated space and air capabilities are entirely new capabilities much more powerful than ASAT capabilities. They are much powerful capabilities in this space era.

In Mr. Austin's idea, the space and the air are separated. The U.S. has its satellite systems in space that provide GPS, precision guidance, reconnaissance and intelligence services to not only the military but also civilians. The services are indispensable for U.S. military on Earth including army, navy and air force, but the systems do not rely on the military on Earth.

The satellite systems are attacked by ASAT weapons, which are in fact space weapons that U.S. air force do not have and are unable to deal with; therefore, U.S. air force are unable to defend the satellite systems it relies on. It has to use U.S. space force to defend its satellites but the U.S. has not established such a force yet and does not seem to have a plan to establish such a force.

On the other hand U.S. satellite systems, though provide vital services to U.S. air force, are unable to defend U.S. air force. Air battle has to be fought with traditional weapons used by aircrafts and air-defense systems. Without such weapons, the satellite systems in space cannot help U.S. air force win an air battle no matter how wonderful the satellite services they provide.

Obviously, U.S. current strategy does not integrate its space capabilities with its air capabilities.

U.S. Air Force Is Not in Danger Even if U.S. Satellites Fail to Function

Suppose all U.S. satellites have been neutralized by China, but U.S. has maintained an air force supported by satisfactory radar, air-defense and anti-missile systems of its army and navy. The U.S. will not lose its Air-Sea Battle in areas near China even if China has a strong land-based air force.

In areas far away from China, China is by no means a rival to U.S. Air-Sea capabilities unless it has built up an aircraft carrier fleet superior to U.S. It takes a few more years for China to make its first aircraft carrier the *Liaoning* combat ready. Building up a comparable fleet will take at lease 2 decades.

In fact, as the U.S. keeps on developing and improving its aircraft carrier fleet, China perhaps will never catch up with the U.S. in that respect.

However, the Air-Sea Battle is a strategy of mid 20th century. The progress of science and technology has enabled the development of a brand new strategy superior to that of integrated Air-Sea Battle.

That is China's new integrated space and sea strategy. However, as mentioned above, we really do not know exactly what China's integrated space and air capabilities are. The description of the war between China and the U.S. at the beginning of this book is only this writer's imagination. It can become a true situation as it is reasonable and achievable in the future, but we really do not know whether the capabilities China will develop are as powerful or even more powerful.

The following chapters describe what has been known about China's efforts in developing the capabilities and the reasons why China is making such great efforts to surpass the U.S. militarily.

The U.S. Shall Be on Its Strict Guard Not to Underestimate China

The most important lesson the U.S. has to learn is the Korean War.

When the U.S. force would soon cross the 38th parallel, Chinese leader Zhou Enlai first strongly warned publicly on September 30: "The Chinese people... will not supinely tolerate seeing their neighbors being savagely invaded by the imperialists." Two days later, as China had no diplomatic relations with the U.S. at that time, Zhou formally notified through the Indian ambassador, that if the American troops entered North Korea, China would intervene in the war.

However General MacArthur and the political leaders in Washington regarded his warnings as political blackmail.

Their underestimate of China's determination and strength led to their bitter defeat soon after Chinese troops had entered Korea.

If General MacArthur had been on his strict guard not to underestimate China, he would have not made rash advance but told his troops to be prepared for Chinese attack. With control of sky and sea and superior weapons, the U.S. could have easily kept most of the North Korea it had taken before entry of Chinese troops.

Now, the U.S. shall be on stricter guard against underestimating China as China is much stronger now so that the consequence of underestimate will be even more serious. However, it seems that some Americans do not.

There is a prestigious Western military expert Ian Easton, who published an article on The Diplomat on January 31, 2014 titled "China's Deceptively Weak (and Dangerous) Military" based on outdated information.

The most absurd is his allegation in the article: "Not to be outdone by the conventional army, China's powerful strategic rocket troops, the Second Artillery Force, still uses cavalry units to patrol its sprawling missile bases deep within China's vast interior. Why? Because it doesn't have any helicopters."

China is now producing a few types of helicopters. Since 2009, China has conducted series production of WZ-10 armed helicopters and delivered them to its army. It is believed that by the end of 2013 there were 60 to 100 such helicopters in service. There has been a report that claims that WZ-10 is better than U.S. AH-64D Apache. (See the section with the subhead "China's WZ-10 Armed Helicopter Better than AH-64D Apache" in Chapter 14).

Moreover, it is common knowledge that China's strategic rocket troops station in 5,000km tunnels with mobile ICBMs instead of their old sprawling bases deep within China's vast interior.

It is said that previously, Mr. Ian Easton was a China analyst at the Center for Naval Analyses in Japan. I hope Japanese decision makers do not fight a war with China based on his advices.

The Philippines underestimated China's determination and overestimated U.S. support when it began the Scarborough Shoal standoff. Before the standoff, the Philippines and China both patrolled the shoal, now the shoal is entirely controlled by China.

5. China Surpassing U.S. in Star War Capabilities

Better Global Positioning System than U.S. GPS

China's Beidou Navigation Satellite System (BDS) has been providing better services than U.S. GPS for the Asia-Pacific region since December 2012. By 2014, BDS has been able to provide positioning data between longitude 55°E to 180°E and from latitude 55°S to 55°N.

It cannot provide global coverage because only 16 of its planned 35 satellites have been launched. However, the services provided in the limited area prove that it is superior to U.S. GPS.

First, it has better accuracy. Its civil services have similar accuracy to GPS, but the positioning accuracy of its exclusive military service for China and Pakistan is said to be 10 centimeters.

The U.S. GPS has only 32 satellites all in medium earth orbits with no geostationary ones. China's BDS has 35 satellites, 5 of which in geostationary orbits, 27 in medium earth orbits and 3 in inclined geosynchronous orbit. No wonder BDS has better accuracy.

There has been the plan to increase the number of satellites to more than 75. In that case, BDS will be far superior to GPS.

Second, BDS provides message services so that one user can send its position and status to another.

Third, it can better resist interference. GPS is vulnerable to interference as proved by Iran's capture of a U.S. reconnaissance drone.

Due to the communications between the BDS user controlling a drone and the drone, interference can be discovered; therefore, BDS can better resist interference.

China has successfully developed chips and terminal products for BDS and is conducting mass production of them. When all the 35 satellites have been put into orbits, BDS will provide global services better than GPS. It is expected that by the time when BDS provides normal global services, it will grab from the U.S. a market share of 80% in China and 20% in the world. That will greatly reduce U.S. government's income from its GPS services and make BDS Chinese government's cash cow.

That is not most troublesome for the U.S. The most troublesome is China's great potential in developing most advanced technology.

In 2003, China wanted to join Europe's Galileo System. It signed an agreement with Europe in 2004 for that purpose. However, Europe failed to see China's potential and allowed China to play only an insignificant role. China withdrew from Galileo System and began to compete with Europe by its BDS in 2008.

Now the fast development of BDS proves that China is a winner in the competition.

It seems that the more China is looked down, the better achievements China can make.

Another example is China's development of its AEW&C aircraft. Previously, Israel signed a contract with China to provide it with Israel's AEW&C, but was force by the U.S. to rescind the contract. However, now China has developed AEW&C aircrafts better that the one Israel wanted but was forced not to sell China.

We have mentioned ASAT and ASAT defense weapons and the part of China's BDS positioning system that surpass the U.S. Those are quite impressive achievements China has made since it adopted the integrated space and air strategy in 2004. They have caused analysts with insight to worry that the U.S. will lose its superiority in a war with China when its GPS, reconnaissance, intelligence and communications satellite systems are damaged by Chinese ASAT weapons.

As pointed out in the last section of Chapter 2, U.S. air force is not in danger if its stealth bombers and fighter jets, AEW&C, carrier-based aircrafts, aircraft carriers and warships remain better than or at least as good as Chinese ones and U.S. has also used its own ASAT weapons to have neutralized Chinese satellites.

Aircraft Carriers v. Space and Air Capabilities

U.S. is strong in its huge aircraft carrier fleet, but as far back as in the 1960's, Soviet military theorists pointed out that the development of missiles is putting an end to the era of large warships. However, their theory did not materialize at that time as it was based on the use of nuclear missile, which may give rise to a large-scale nuclear war that neither the Soviet Union nor the U.S. wanted to fight.

However, due to development of science and technology, the new anti-warship missiles deployed by China constitute serious threat to aircraft carriers' survival. A hypersonic cruise missile can be controlled to fly along unpredictable projectile and hit its target accurately. That is what China gives priority to in its military modernization.

The U.S. always worries about China's DF-21D missiles, but very probably,

DF-21Ds are not the real aircraft killers but decoys to attract U.S. fleet's missile-defense firepower. The real killers may probably be accurate cruise missiles. A J-20 stealth fighter jet can fire cruise missiles to attack an aircraft carrier. The technology now can make anti-ship missiles hit target far away at high speed without being detected. It is now impossible to protect a large surface warship from being hit by missiles especially near Chinese coast where there are large number of missiles. The U.S., therefore, has problems in attacking China. Still, China does not dare to fight a war against the U.S. because U.S. navy dominates the oceans and may easily cut China's trade lifelines.

Can China develop a comparable aircraft carrier fleet to counter U.S. dominance?

Due to lack of the expertise, technology and experience in designing, building and operating aircraft carriers, if China follows U.S. strategy to develop its aircraft carrier fleets as strong as the U.S., it takes at least 20 years to build and 10 years to learn the skill to operate China's aircraft carrier fleets while the U.S. will make much improvement in its fleets for China to catch up, how can China become powerful enough to prevent being bullied by the U.S., especially its trade lifelines from being cut?

However, what if China develops instead a fleet of aerospace bombers able to destroy U.S. navy far away from China? With existing technology, it takes much less time to develop reusable aerospaceplanes able to travel between space and earth. U.S. space shuttle is such an aerospaceplane. Only, it is not large enough.

Large aerospace bombers will make worthless U.S. expertise in designing, building and operating aircraft carriers. Excluding such expertise, in fact, U.S. military is not superior to its Chinese counterpart in developing new technology and tactics. Although the U.S. remains the number one world power, China has huge potential. China has the advantages of having less restriction by its tradition, dogma and the investment previously made. As a result, its plans will probably be more flexible and vigorous.

Moreover, development of aircraft carriers has nothing to do with hypersonic weapons of our space area; therefore, China will have two heavy financial burdens if it develops both aircraft carriers and hypersonic weapons.

However, the development of hypersonic weapons is a byproduct of the development of space technology as they both concern super high speed. Mastery of the technology in achieving and controlling the high speed of spacecrafts is close to mastery of hypersonic weapons. The additional investment will be much smaller compared with the establishment of a huge aircraft carrier fleet rival to America's.

Therefore, the only way out for China is to develop integrated space and air capabilities for both attack and defense.

The vital part of the capabilities is not ASAT or ASAT defense weapons or satellite global positioning system, but the manned attack aerospaceplanes described at the beginning of this book. Those aerospaceplanes will be a vital part of China's space program.

Military Potential of China's Space Program

China is making great efforts to carry out its ambitious space program. There is speculation that it may establish a military base on the moon. However, that is indeed not practical as attacking earth targets from the moon will be very much complicated.

What shall be worried is the attack of earth targets from a space station. As the missile in space is already moving at high speed, there is no need for much fuel like an ICBM or SLBM. The missile will be much smaller and lighter.

If China were allowed to join the international space station, it would be impossible for China to store weapons at the station. However, as China will build its own independent space station, it will be much more difficult for international inspection to prevent China from storing weapons at the station.

Lunar Rover's Glitch Proves Greater Efforts Needed to Get Space-Air Capabilities

However, building and operating a space station is very expensive and complicated. Is China competent to do that?

The breakdown of China's lunar rover after six weeks in operation proves that there is still a long way to go in China's efforts to complete its ambitious space program.

Hong Kong English Newspaper South China Morning Post (SCMP) says in its report "Last-ditch efforts to salvage mission of China's stricken Jade Rabbit lunar rover" on April 18, 2014, "Engineers identify power blockage as cause of Jade Rabbit's breakdown, which has left the rover parked up on lunar surface for two months."

However, in a Beijing interview with Hong Kong newspaper Wenweipo's reporter on April 16, Prof. Ouyang Ziyuan, Chief advisor to China's moon project, said frankly: So far there is still no way to know the exact causes for the breakdown. Some people believe that the lunar rover's moving parts have been jammed by dust on the moon; others blame the malfunction of its electronic components; still others ascribe the glitch to computer failure.

Prof. Ouyang talked much about the difficulties for the next landing on the moon by Chang'e IV. In fact, Chinese scientists have not reached a consensus on where it shall land on the moon. Due to the requirements for safety, the landing site shall have even topography, abundant energy and easy access to telecommunications.

The location where scientists want earnestly to land on shall be the south pole of the moon, but the topography there is too complicated and there will not be enough sunshine. Some scientists suggest that the spaceship shall land twice on the moon. It shall fly from its first landing site to a second one in order to gather more information about the moon.

Anyway, China lags far behind the U.S. in space exploration, but it has been trying hard to catch up while the U.S. is not doing much due to lack of fund or awareness of the strategic importance of space.

Avant-garde Strategy: Integrated Space and Air Capabilities

China has kept strictly confidential its integrated space and air capabilities. As a result, we cannot have any exact idea what capabilities China wants to develop. However, with some knowledge about aerospace technology, we can be very clear that if the U.S. had kept on developing its space shuttles, perhaps, it might have developed an aerospaceplane capable of flying in both space and air instead of the above-mentioned aerospace drone.

Such a shuttle will be flying in its orbit and give the impression that it is a space station, but when there is a war, it will fly down to attack the enemy. As it flies in its orbit at the speed of 7,700 m/s, i.e. Mach 22.6, the missiles fired by it will have the same initial speed as super hypersonic weapons. It is difficult to develop any defense against such missiles.

I believe an aerospaceplane like the above mentioned shuttle is precisely what China is trying to develop with a little success now and will continue to develop along with its lunar program

However, the U.S. does not seem to have realized space shuttle's military potential. It has failed to make any follow-up efforts in developing large manned aerospaceplanes on the basis its space shuttles. Instead, it is developing an unmanned aerospaceplane much smaller and is satisfied with its success.

Chinese strategists, however, saw the potential of aerospaceplanes long ago and have succeeded in persuading Chinese leaders to accept their plan for the development of integrated space and air capabilities. The most important part of such capabilities, I believe, is huge attack aerospaceplanes.

For a nuclear war, the aerospaceplane can carry out nuclear first and second strikes. Its nuclear missiles are more powerful than ICBMs and SLBMs as they fly at the hypersonic speed of Mach 22.

For a conventional war, as described in the beginning of this book, an aerospaceplane battle group makes aircraft carrier battle groups obsolete.

U.S. Losing Leadership in Aerospaceplanes

Boeing began to develop X-40 aerospaceplane in late 1990s. NASA took over X-40 and began its X-37 Orbital Test Vehicle (OTV) project on the basis of X-40.

U.S. Department of Defense took the project over in 2004 and successfully conducted the first drop flight test on 7 April 2006 and the first spaceflight test from 22 April to 3 December 2010.

The U.S. intends to use X-37 to rendezvous with satellites to refuel them, or to use its robotic arm to replace failed solar arrays. X-37 can also support space control for defensive and offensive counter-space maneuvers.

The U.S. later conducted two test flights of X-37B, a modified version of X-37, respectively in 2010 and from 2011 to 2012. X-37B remained in orbit for 469 days and successfully landed autonomously on June 16, 2012. It is regarded as a tremendous success and has made the U.S. world leader in the development of reusable unmanned aerospace vehicle.

One more test flight of X-37B began successfully on 11 December 2012. The aerospaceplane remained in orbit by March 2014 more than 470 days after it was launched.

With such tremendous success, the U.S. is sure of its leadership in aerospaceplanes.

China Shocked the U.S. with its Aerospace Fighter Able to Reach the U.S. in One Hour

Chinese President Xi Jinping's call in late April 2014 for the PLA to speed up building an integrated space and air force has drawn keen international attention. Foreign media are publishing sensationalizing reports on that. There is no shortage of bold speculation. Some media have even reported that China is testing an aerospaceplane codenamed Shenlong that is able to reach the U.S. in one hour. It deals a heavy blow at the U.S. if what they report is true.

According to a report that the U.S. has most recently issued, China is testing Shenlong aerospace fighter that is able to reach the U.S. in one hour. The report says that Shenlong is much smaller than U.S. X-37B aerospaceplane and has been researched for much less time, but it can carry small rockets with surprising power.

There is information that Fujian Longxi Bearing (Group) Corp., Ltd. conscientiously cooperated with the research institute for the fighter in producing the bearings it needed. For that the institute awarded the company a board of inscription and a letter of gratitude.

China has successfully carried out test flight of Shenlong, but it is too small. Now Chinese space designers are perhaps trying to develop an enlarged version of Shenlong aerospaceplane.

Chinese aerospace equipment experts have held quite a few secret meetings for that and have begun to work on the design of the large aerospaceplane.

Russian media has also pointed out: A large aerospaceplane is of great practical value; therefore, China has invested lots of funds and workforce in the research and development for it and has made considerable achievements.

Moreover, China is developing other models of aerospaceplanes.

J-28, Another Space-air Fighter that China Is Developing

A Russia media revealed in April 2014: China is doing research for the development of J-28, its 5th-generation fighter jet. It is precisely what China has always been doing in developing its weapons: Production and improvement of one generation, research and development for the next generation, preliminary research for a third generation and exploration for a fourth.

China's Aviation Research Institute No. 611 is now doing the research for J-28, which shall be a 5th-generation multi-function stealth fighter jet. It is said that J-28 is able to conduct accurate conventional hit to counter the nuclear threat from the entire solar system.

The laser weapon J-28 carries is so powerful that it can melt a nuclear missile launched from any planet or satellite in solar system. J-28's speed is so high that even if it fails to intercept a missile the first time, it can chase the missile to hunt it.

In addition, it is such a wonderful stealth fighter that it cannot be detected by either radar or radio telescope.

The report says as the United States has conducted enough survey to know the Mars, it is expected that the U.S. will establish a nuclear missile base on Mars in 2018.

In order to deal with the nuclear threat from solar system, China has begun developing J-28, its 5th-generation multi-function stealth fighter jet.

China Is Building World Largest Radio Telescope to Detect Missiles from Space

China is building the world largest radio telescope with a diameter of 500 meters much larger than the 300 meters of U.S. Arecibo Radio Telescope. It occupies a site of 30 soccer fields. Its construction is expected to be completed in September 2016.

In April 2014, Global Times interviewed Li Di, chief scientist of the Radio Department of the National Astronomical Observatory at Chinese Academy of Sciences.

Li said that the radio telescope project (FAST) began construction in March 2011 and is expected to be completed in five and a half years. The project has been making smooth progress so far but as it is the world's largest, there will be lots of challenges in the course of the construction.

Li explained that the telescope was being built in Pingtang County, Guizhou Province due to the natural pit of the Karst topography there. In addition, as Guizhou is a remote province with a relatively small population, the radio interference from mobile phones, radars and satellites is smaller there.

FAST, when built, will be twice sensitive than U.S. Arecibo and 10 times quicker in searching the space.

As for the question of what is the use of such a huge telescope, Li said it could be used to detect remote signals and matters in universe such as gas, microwave laser and pulse star for better understanding of the ingredients of the matters in universe and the history of the evolution of universe.

However, according to Chinese leader Xi Jinping's recent call for developing integrated space and air capability for both attack and defense, China is making efforts for space-air battle. The huge radio telescope can also be used to detect missiles and space vehicles coming from space to attack China. In fact, the radio telescope can easily be turned into huge space radar.

Speculation about China's Strategic Bomber

Foreign media have much speculation about China's plan to build its stealth strategic bomber. Due to China's habit to keep its weapon development confidential, the speculation is often groundless guessing as information about such top secret is utterly unavailable.

However, as China has been conducting discussions with Russia on joint development of large aircrafts, Chinese officials taking part in the discussions may have said something about large strategic bomber to draw out some information from Russian experts on large aircrafts, but has thus also revealed China's interest in large strategic bombers. Therefore, I believe Russian sources are more reliable.

On July 16, 2013, Voice of Russia published a report on Russia's plan to develop its strategic bomber, in which it mentioned China's plan to build its strategic bomber.

Huanqiu.com soon published a report based on Voice of Russia' report. It said: like U.S. B-2, Russia's bomber will be maximal subsonic with major characteristics of long range and small probability of being detected by radar. It is different from China's future strategic bomber in this respect. China's strategic bomber will be supersonic. According to analysts, China has to spend a lot of funds in implementing its plan for such a bomber as the technology will be very complicated.

A factor of special concern is that unlike the United States and Russia, China has no experience in that respect. In fact, if China wants to put such a bomber into mass production, it has to put in more resources than it has put in the projects of developing two 4th-generation fighter jets. Perhaps, it has to incur a cost exceeding its entire manned space program.

Chinese Air Senior Colonel Confirms Development of Strategic Stealth Bomber

On October 17, 2013, in an interview with people.com.cn, Senior Colonel Wu Guohui, associate profession of National Defense University and a super pilot, bared to web users the secret of air battle system. He said in the interview that various countries have attached importance to stealth strategic bombers again. In the future, China would also develop a new long-range strategic bomber.

Wu Guohui said that for a period of time in the past, no importance had been attached to the development of bombers. The generally accepted idea is that such a bomber is heavy with poor maneuverability so that it is easy to be shot down by fighter jets and ground fire. However, at present, the U.S., Russia and China have all attached importance to strategic bomber with stealth capability.

What Wu said was but the world trend that China has to follow. As he is not an insider to China's top secret plan on such a bomber, his words fail to unravel the mystery.

China's Efforts in Developing Long-Range Bombers

The capability of a country's military aircrafts is a country's military secret; therefore, in spite of much speculation at home and abroad, PLA has kept on guarding such secrets closely. However, since PLA has been making great efforts in developing bombers, some information will unavoidably leak out.

In late April 2014, there were exposures on the Internet of PLA's super weapon H-8 bomber similar to U.S. B2. According to Jane's Defense weekly, China is developing H-8 stealth bomber that can carry stealth cruise missiles with capability similar to U.S. B2A.

According to speculation it can carry 3 supersonic anti-warship missiles, 4 air-to-ground missiles and 6 medium- and short-range air-to-air missiles able to be a rival to U.S. B2A. It can break air defense at a speed of Mach 1.2. With a range longer than 10,000 km and ability of refueling, it can carry new stealth cruise missile to attack North America.

There has been report that maiden flight may be conducted soon as China finalized the design of its 4 Taihang engines in 2008.

However, Pinkov, the founder of Canada's Kanwa Defense Review said in an interview that China was not able to develop stealth bomber due to lack of technology for powerful engine, stealth materials and coating.

Moreover, a major general (Mathews) in charge of projects at U.S. Air Force Operations Command believes that as the capability of a stealth bomber in bringing the fiercest fire power above the enemy's head at an instant can be replaced by weapons launched outside of defense area, F-22 stealth fighter, attack drone and hypersonic missile, Russia and other air powers simply have no plans to build any strategic stealth bomber.

China's New H-18 Stealth Fighter-Bomber Able to Reach Guam

In November 2013, U.S. militaryphoto website carries a report on China's plan to build H-18 stealth bomber, which is a stealth fighter-bomber instead of a strategic bomber.

The bomber is 28 to 30 meters long, much smaller than U.S. B-1B's 44 meters length.

According to foreign military fans' speculation, H-18 has a range of 8,000 to 9,000 km and maximum combat radius of 3,500 to 3,700 km and a speed of Mach 2. It can carry a load of weapons of 12 to 15 tons.

It has an inside weapon cabin 8 meters long for 72 100kg Leishi small precision-guided bombs or 4 Changjian-10A cruise missiles that may carry out tactical or nuclear attack at Guam. It may also carry 4 Yingji-12 supersonic anti-ship missiles or 4 Jingji-100 super long-range anti-ship missiles to kill an aircraft carrier.

It has adopted lots of stealth technology to become world's first stealth fighter-bomber and stealth supersonic bomber. The H-8, if developed, may serve as a stealth strategic bomber able to attack the U.S., but it has not yet conducted its maiden test flight. In addition, there has been no official confirmation of any plan to develop it.

What we have official information is China's H-6K bomber.

In October 2013, China made public for the first time its H-6K bomber that carries 6 Changjian-10 (CJ-10) air-to-ground cruise missiles. In addition, H-6K can carry CJ-20 cruise missile with nuclear warhead that China is developing.

However, military analysts believe H-6K is but China's transitional bomber and China is developing a strategic bomber with a range of 7,500 miles (perhaps, they mean H-8).

Like China's most new weapons, there has been no official information about H-18.

China's Plan to Develop Strategic Bomber as Revealed by Russian Media

On July 16, 2013, Voice of Russian reported that Russia would begin technical design of its new strategic bomber in 2014.

It said that like U.S. B-2, Russia's bomber will be maximal subsonic with major characteristics of long range and small probability of being detected by radar. It is different from China's future strategic bomber in this respect. China's warplane will be supersonic.

According to analysts, China has to pay a lot in implementing its plan for the warplane and the technology will be very complicated. A factor of special concern is that unlike the United States and Russia, China has no experience in that respect.

In fact, if China wants to put such a project into mass production, it has to put in more resources than it has put in the projects of developing two 4th-generation fighter jets. Perhaps, it has to incur a cost exceeding its entire manned space program.

At that time, except for the supersonic speed, no details whatever have been given about China's strategic bomber.

Huge Super Nuclear Bomber Carrying 200 Nuclear Bombs

On October 29, 2013, China's qianzhan.com finally got some more information about China's strategic bomber. It said in its report titled "Shocking Revelation of China's huge nuclear bomber that carries over 200 sets of missiles" that according to Russian media report, the PLA is developing a new generation of strategic bomber. Due to the breakthrough in the technology to reduce the size of high-temperature air-cooled nuclear reactor, China's new bomber will not only be nuclear powered but also be armed with shocking weaponry due to its great loading capacity.

It is said that the bomber codenamed H-11 surpasses China's existing H-6K medium-sized bomber by far and will be what U.S. military fears. The cruise missiles carried by H-6K bomber may be a threat to U.S. homeland, while the huge bomber H-11, if produced, will be an unthinkable threat to the U.S.

The report says that Chinese military has improved H-6 to make H-6K a transitional bomber to satisfy present needs. That is why the improvement does not touch its aerodynamic design. As one of the three nuclear striking means, strategic bomber has its unique role and position and is a political means indispensable for a political power. That is why as a country pursuing the Chinese dream, including the dream for a militarily powerful China, China has to develop a new generation of bomber with much greater lethal power.

China began development of strategic bombers in the 1970s, but lacked the technology for any breakthrough. Both its H-6I and H-8 strategic bombers failed to achieve its goal. At that time, China lacked the technology in each and every aspect of a strategic bomber. Through accumulation of technology for three decades, China now is entirely capable of developing a strategic bomber. However, it takes a long time to design and commission the bomber and make it combat effective.

In particular, China has no experience in the research and development for such a bomber, especially its electronic equipment and engine. A problem in any part of the complicated project may cause some delay. Even if China has developed a strategic bomber, it has no experience in operating it. It has to start from nothing in training the tactics and skill in using the bomber, coordinating with other aircrafts and ground and air supporting services, maintaining the bomber, etc. Analysts believe the research and development for H-6K have not only provided a transitional aircraft to satisfy current needs, but were also technological and theoretical research for operating the new strategic bomber.

In light of the previous generation of Russian and U.S. bombers, the H-6 bombers in service in Chinese air force should be regarded as a medium-sized bomber. It can carry 6 CJ-10 cruise missiles outside its cabin. The number of short-range missiles it can carry depend on the number of underwing pylons and the capacity of its weapon bay.

The next generation of large Russian bomber is required to carry Kh-555 cruise missiles with a range of 2,000 km and Kh-101/102 cruise missiles being developed. The two kinds of cruise missiles are all able to carry nuclear warheads with a range exceeding 2,500 km. All the cruise missiles have to be hidden in weapon bay to enhance undetectability and the ability to break air defense. They are much more advanced than H-6K.

The new nuclear powered strategic long-range bomber being developed by China codenamed "Light of Five Stars" will enhance China's nuclear deterrence to unprecedented high and bring about a revolutionary change in the balance of air force in the world. It will enable China to have overwhelming air superiority. Its cruise speed will be Mach 3.6 and it can remain flying at high speed for 3.5 months incessantly. It has a capacity to carry 170 to 210 sets of nuclear bombs depending on the density and scale of its targets.

Suspected Photo of China's Strategic Bomber

U.S. "Foreign Policy" magazine reported on December 30, 2013 that a set of photos of bombers posted by a Chinese military forum were crazily reposted on Chinese Internet. There are photos of double-engine H-6K stealth bombers, but as there is the photo of a bomber like U.S. B-2 on Chinese runway, analysts believe that H-6K is not the only stealth bomber China is developing.

It gives rise to the speculation that China is developing a new highly sophisticated stealth bomber to quickly enhance its military strength.

However some journalist, though interested, questioned whether the photo has been edited.

The Huge Aerospace Bomber Described at the Beginning of This Book

China's strategic aerospaceplane project is a closely guarded secret. People in Russian military industry have the speculation that China wants to develop a huge nuclear powered strategic bomber because China and Russia have been discussing joint development of long-range aircrafts since 2012. The Chinese party to the discussion may have mentioned the desire for a huge long-range strategic bomber that may remain flying for 3.5 months.

Russian military industrial circle only have some vague information about the bomber, but due to their conventional thinking, they believe that since the aircraft has to stay for 3.5 months in sky, it has to be nuclear powered.

In fact, from an integrated space-air point of view, there is no need to develop sophisticated nuclear engine for the aircraft, a rocket engine using current technology can do the job.

Like U.S. space shuttle, a huge rocket can send the huge bomber to its low obit in space, say an orbit 100 km high. If the bomber is too big for one rocket, it can be sent to space piece by piece with its crew in the pieces. The crew will assemble the pieces into the aerospaceplane in space. That can be done with existing technology.

When the aerospaceplane has been assembled in space and begun to operate like a space station, transport rockets will send fuel, weapons and other supplies to the aerospaceplane.

In fact, the aerospaceplane needs not be very big as it keeps flying in its orbit by inertia. There is no need for a heavy engine for propulsion. It needs fuel only when it has to go down to attack or change orbit to avoid being attacked. The research for such a spacecraft mainly concerns the way to control the spacecraft and its weapons to ensure accuracy of its strikes.

Landing is certainly a big problem, but it can remain in orbit for a long time and have its crewed replaced by transport rockets. It can also fly to a huge master space station for refueling and getting supplies and replacement of crew, equipment and weapons.

If a lunar base has been established, the spacecraft can land on the moon for maintenance. That will be much easier as there is no atmosphere and much smaller gravity on the moon.

It remains flying in its orbit by inertia so that it needs no engine to maintain high speed. Unless it turns on its rocket engine to change orbit or go down to attack, it has no infrared emission so that good shape and coating alone may make it undetectable.

Such a spacecraft is much more powerful than an aircraft carrier battle group. It takes four days for a group to reach its target area 5,000 km away at top speed of 30 knots while it takes less than 15 minutes for the spacecraft to reach there.

For China, development of such a spacecraft will take less time than development of a most advanced Ford class aircraft carrier and will perhaps cost much less. A Ford-class aircraft carrier costs $11.3 billion. I don't think a huge attack aerospaceplane costs so much. Even if it costs twice as much, it is worthwhile as it can easily kill two carrier battle groups in minutes.

The question now is not whether China is able to develop the technology for the spacecraft but whether China really has a plan to develop a huge strategic stealth bomber, the weapon really has integrated space and air capabilities in our space era.

As described in the beginning of the book, one such spacecraft is more than enough to kill two aircraft carrier battle groups. A country with four such spacecrafts in orbit, two carrying conventional and two, nuclear weapons will be enough to dominate the sky and space.

When there are such spacecrafts, aircraft carriers are obsolete.

Secrets of the Space Weapons China Will Develop

On June 18, a major think tank of Chinese military, the National Defense Policy Research Institute of PLA's Academy of Military Sciences, released its report "Strategic Assessment 2013", which devotes one of its six chapters to the international competition in space.

The report holds that since the beginning of the 21st century, China has been faced with increasingly severe security threat from the space.

In the new century, all major countries in the world regard space as an area for them to enhance their comprehensive power and an important base of their military's strategic capabilities; therefore, they have all formulated and readjusted their national space strategies. As the number of countries exploiting space resources grows and their scope and extent of exploitation widens and deepens, the strategic importance of space has risen to an unprecedented high. Major countries in the world have speeded up the formulation and readjustment of their space strategies in order to obtain preemptive advantages. Due to rising strategic demand and advances in space technology, major countries and regions in the world have kept on increasing their investment in space capabilities and doing their best to achieve preemptive advantages in space. As a result, the situation of strategic contention in space becomes complicated and space arms control, very difficult.

The report points out that as space technology keeps on developing and its application extends, space military strength, especially space information systems, play increasingly prominent role in modern local wars, military conflicts and development of armed force. There has been information that more than 60 countries and regions have made military application of space technology, especially in collecting intelligence. Military superpowers rely greatly on their space systems. World people have a common desire now to oppose militarization of space and safeguard space security and order.

The report analyzes the severe threat from space to China's security and points out the three major areas of threat:

1. The threat of space reconnaissance and surveillance. Over the past few years, some countries and regions have improved their space reconnaissance equipment. There have been continuous emergence of various new types of satellites for electronic reconnaissance, optical and radar-imaging, marine surveillance and early warnings on missile launches. Relevant countries collect China's military, political, economic and technological intelligence by means of all weather, near real-time, multispectral and high-precision photography, mapping and surveillance.

2. The danger of restriction and deprivation of freedom of actions in space. A few countries are stepping up their research and development of space offensive weapons, setting up space combat forces, issuing space war doctrines as well as organizing space offensive and defensive drills in their attempt to interfere, blind and destroy when necessary enemy's space platforms, effective loads and space information chains. A small number of countries pursue absolute superiority in space, adopt a policy to control the space and take vigorous precautions against so-called "Chinese space threat". Their intention and ability to strive to conduct space confrontation and deprive opponents' freedom of activities in space shall not be underestimated.

3, The threat of attack from space to ground targets. Space powers have actively developed space-to-land, space-to-sea and space-to-air attack capabilities using ballistic missiles, near-space aircraft, spacecraft, space-atmospheric vehicles, etc. They are speeding the establishment of their systems of space-to-ground offensive capabilities. As a result, China's important strategic facilities face serious threat.

The report points out China's task now and in the future to persist in advocating peaceful utilization of the outer space, resolutely oppose along with other countries militarization of space and space hegemony and promote the establishment of rational international space new order.

What China wants to develop for its integrated space and air capabilities has remained a well-guarded secret, but now the report gives us some clues:

In fact, we can precisely learn from the report the areas China is making great efforts to catch up and even surpass the United States.

1. At the beginning of this chapter, we described where China's Beidou Satellite Navigation System surpasses U.S. GPS. It proves that China does not lag behind the U.S. in reconnaissance and surveillance satellites. China ranked the first in the world in launching 20 and 19 satellites respectively in 2012 and 2013, of which 97% are China's own satellites. Knowing well the trend of development in reconnaissance and surveillance technology, China certainly is making great efforts to develop such technology. It seems to me that the U.S. is instead facing the threat from China.

2. The same is the case for the second area. China's ASAT and anti-ASAT capabilities are much better than those of the United States. There are quite detailed descriptions of them in Chapter 4.

On January 14, 2013, Reuters mentioned in its report "China's space activities raising U.S. satellite security concerns" a classified U.S. intelligence assessment that warned about the "growing vulnerability of U.S. satellites after "analyzing China's increasing activities in space".

3. As for space attack capabilities, I will describe in this chapter, China surpasses the U.S. in HGV (in a section titled "China Tests Mach 10 Hypersonic Weapon" in Chapter 7) and has almost caught up with the U.S. in aerospaceplanes. (Refer to the section "China Shocked the U.S. with its Aerospace Fighter Able to Reach the U.S. in One Hour" in this chapter.)

China has been vigorously developing the space weapons in the three areas the think tank regards as the threat that China is faced with. The progress China has made in those areas has already made the U.S. uneasy.

As for what China attaches the greatest importance to in the above-mentioned three areas, I believe as the integrated space and air capabilities are for both attack and defense, China will make the greatest efforts in developing ballistic missiles, near-space aircraft, spacecraft, atmospheric vehicles and other new space weapons for space-to-land, space-to-sea and space-to-air attack.

PART 2 THE ARMS RACE

6. China's Arms Race with the U.S.

China Relied on the U.S. for Protection of Its Trade Lifelines

As China has grown into an industrial country that relies heavily on foreign trade, protecting its trade lifelines become vital for it. If its shipping routes are blocked, not only it will not be able to send its goods abroad, it will have no access to the oil and other resources it needed.

As its navy is negligible compared to U.S. navy, it relies on the U.S. to protect its trade lifelines; therefore, like a girl courting a handsome and wealthy man, China has done almost everything to please the U.S.

First, in spite of China's insistence on resorting to force when Taiwan announces independence, China has done the best to win over Taiwan peacefully as the U.S. wants a peaceful solution of the Taiwan issue.

Second, China avoids persecution of the dissidents the U.S. openly supports and even allows them to leave for the U.S.

Chen Guangcheng is the best example. When Chen has obtained U.S. protection after he fled into U.S. embassy in Beijing, China quickly allowed Chen and his wife and children to move to the United States.

Third, China is willing to incur heave losses to please the United States.

The most obvious example was China's support for American actions in overthrowing Libyan dictator Gaddafi in spite of the heavy losses China might suffer due to that. It had withdrawn 50,000 Chinese working there and suffered heavy losses as its projects in Libya could not go on when Gaddafi had been removed.

The U.S., however, regards what China has done as something China ought to do so that it does not appreciate China's efforts to please it.

On the contrary, as Chinese economy kept on growing at high speed, the U.S. began to be afraid that one day China might surpass the U.S. and replace the U.S. as the greatest superpower. It, therefore, wanted to contain China by its pivot to Asia.

Upset by U.S. Containment, China Began Its Arms Race with the U.S.

In July 2010, China reacted angrily at statements concerning South China Sea disputes by U.S. Secretary of State Hillary Rodham Clinton during an ASEAN security gathering. China began to quicken the development of its navy, but at that time, its arms race with the U.S. did not begin in earnest.

Anyway, in addition to the busy testing of its aircraft carrier, it conducted trial operation of its Beidou satellite GPS system and launched its third 071 landing platform dock (LPD) in September 2011.

071 LPD is a 20,000-ton amphibious transport dock similar to the U.S.-built San Antonio-class LPD. With a range of 11,000 km and a capacity to carry 500-800 troops, 15-20 armored vehicles and 6 helicopters, it will be quite useful if China has to solve South China Sea disputes by force.

Since President Obama's high-profile announcement of America's return to Asia in November 2011, China has decided to counter military threat with military strength, but ostensibly, it responded with a Xinhua comment entitled "Constructive U.S. role in Asia-Pacific welcome" and reiteration of its statement that China's peaceful rise constitutes no threat at all to America.

Sun Tzu's *The Art of War* gives the advice: "The art of war is an art of deception. Hence, one shall make his enemy believe that he is unable to fight when he is able to and that he has no desire to fight when he is going to." Having followed Sun Tzu's teachings for over 2,000 years, China believes that America's activities in providing arms to and holding joint military drills with China's border dispute opponents are aimed at containing China and that America's statements to the contrary are but deception.

China launched its first 071 LPD in December 2006 and its second 071 LPD 4 years later in November 2010, but it launched the third 10 months later. In addition to putting its own satellite GPS system into trial operation, it declared its plan to launch 6 more satellites in 2012 to improve the system. Guided by this system, China's missiles and bombs will be much more accurate. Why was the haste? Because China was conducting an arms race with America in earnest. It now wants to catch up with and surpass America in military strength as soon as possible because it believes that America wants to bully it.

There is something quite similar between the China now and the Germany after World War I. German people suffered greatly when German economy collapsed due to the heavy indemnity imposed on it after it lost World War I. Chinese people suffered greatly when Britain defeated China and forced it to accept import of opium. The misery lasted for over a century afterwards as China was bullied by various foreign powers and finally invaded by Japan for fourteen years. Hitler became popular as German people wanted revenge, but Chinese people do not want revenge. They want to put an end to its century-long history of being bullied by foreign powers.

Does America want to bully China? Certainly not. America is uncertain whether China's rise will be a threat to its security and interest and world peace or beneficial to the world. It switched its strategic priority to China merely for the purpose that it can adopt preemptive action in case China turns out to be a threat. However, this shows America's dire lack of understanding of China.

First, America is ignorant of Chinese people's sensitiveness to bully by any foreign power. U.S. deviation from its traditional attitude of avoiding being involved in border disputes between other countries gives Chinese people the impression that it is playing off China's neighbors against China in order to profit from the border disputes. Chinese people have begun to regard America as their enemy.

Second, it lacks basic knowledge about Chinese history, which is a history of dynasties. Usually, a major dynasty remained prosperous for about one century after its establishment. The CCP (Chinese Communist Party) Dynasty has been established for only three decades since Hua Guofeng, the successor of Mao Dynasty, was dethroned; therefore, there will be at least several more decades of prosperity in the future for China. People kept predicting hard landing of Chinese economy, but it has never occurred because Chinese leader and his talented assistants with moral integrity are too wise and clever to allow it to occur.

People believe that it is safe to predict China's tremendous growth cannot last for more than 30 years because Japan's rapid growth lasted for only 30 years in the past. However the figure 30 has no magic force for China as China is an entirely different country.

My greatest worry is that a despot like Mao may be selected as a successor to the core of the collective leadership (paramount leader) in China's current system of succession. When such a despot comes to power, he may exploit the enmity among Chinese people to confront America when China has grown stronger than America. Such confrontation may have disastrous consequence to China and the world but in China's political system, no one can stop the despot.

Therefore, I hope America realizes the above-mentioned danger and strives to make it as strong a rival to China to maintain the balance of strength. That requires it to conduct economic reform to give play to American people's talents and diligence as well as a political reform to overcome the lack of cooperation between the two major political parties.

Especially, the U.S. shall not fail to see China's potential to surpass it militarily. First, China has a leader who knows well not only military strategy but also the technology required in modernizing Chinese military. Second, Chinese military has an unlimited budget. Third, it can get advanced technology from Europe, not only from Ukraine and Russia but also from France, Germany, Britain and other EU member countries. The last but not the least, China has talented scientists and engineers who develop weapons for China with patriotic dedication.

Chinese Military's Unlimited Budget

On March 7, Isaac Stone Fish, a well experienced journalist on China expressed his frustration about the mysteries of Chinese military in his article "The Black Box of China's Military".

Such frustration is common among people not only outside but also inside China.

In my book *Tiananmen's Tremendous Achievements Expanded 2nd Edition* released to mark the 25th anniversary of Tiananmen Protests, there is the section titled "Looking through the Black Box", excerpts of which are given below:

> There is a way to look through the black box. We can judge the events taking place in the black box in secret by the circumstances and happenings before and after an event.
>
> In fact, keeping secret information about power center is not merely CCP's modern practice. It was also a practice in feudal courts. Still we Chinese have been able to write thousands of years of Chinese history in details including some real mysteries.
>
> We have the well-known saying about palace coups "Sound of axe and shadows in candlelight, eternal mystery." It is a saying about palace coups originated from the abnormal death of Emperor Taizu (927-976) of the Song Dynasty (960-1279) and his succession by his brother instead of his son. People heard sound of axe and saw shadows in candlelight the night the emperor suddenly died when only the emperor and his brother were in the room.
>
> The murder is certainly not found in official historical records as no one dared to accuse the emperor who succeeded Emperor Taizu, of murder. The murder was revealed in unofficial history that records what happened before and after the emperor's death.
>
> Upon comprehensive logical analysis of Emperor Taizu's health conditions before his death, his young age of 50, his unexpected sudden death, the abnormal succession by his

brother and the abnormal death of Emperor Taizu's sons under Emperor Taizu's brother's reign, we can safely believe that Emperor Taizu was murdered by his brother.

We make comprehensive logical analysis of all the circumstances and happenings before and after the event and people's characters and will thus be able to break the mystery in the black box. It may not be 100% accurate, but it is at least reasonable, logical and believable.

Certainly, I hope there will be later memoirs to provide accurate information about the event, but the probability of the emergence of a reliable relevant memoir is small. Therefore, without such relevant memoir to provide a true record of the event, we have to satisfy with rational, logical and believable analysis based on facts. However, even without such a memoir sometimes, such rational, logical and believable speculation based on facts may be proved by later development.

As mentioned above, what I said about Xi Jinping being selected by Jiang Zemin was speculation, but it was based on reasonable and logical analysis. Now, it has been proved by Jiang's support for Xi with his decision on punishing Bo Xilai harshly and his selection of five protégés as PSC members who will retire at the next CCP congress so as to allow Xi to choose their replacements in 2017 and by what Xi has done since he took over.

Note: Since Jiang's retirement, he has always controlled the PSC (Politburo Standing Committee) through his protégés who have a majority in the PSC. At the end of each five-year term of the PSC, there are always some of Jiang's protégés who have not reached the age of retirement and will continue to be members in the next PSC. This time is exceptional. All his protégés in the PSC will retire when the term of the PSC ends in 2017.

Using this method, we can see through the black box of Chinese military and answer the two questions raised by Isaac Stone Fish in his article: How much is China's military budget? And to what extent does Xi Jinping control Chinese military?

First, we shall see that China is an autocracy instead of a democracy like the United States. How much Chinese military spends does not require the approval of the NPC (National People's Congress, China's parliament). According to Chinese Constitution, NPC has supreme power, but in fact, it is but a rubber stamp that will approve everything the CCP (Chinese Communist Party) tells it to approve.

Therefore, the military budget approved by NPC does not restrict Chinese military spending like U.S. budget does. The budget figure is a carefully determined figure to please Chinese people and ease other countries' concerns.

How much is Chinese military actually allowed to spend?

There is no limit! It can spend as much as it needs and China can afford.

Does this mean that it can spend at will? No, it has to obtain approval from the top leader before it can get funds from China's exchequer.

Who is the top leader that controls both the exchequer and the Chinese military? Is he Xi Jinping? No. It is the core (the term used by Deng Xiaoping) or paramount leader (the term used outside China) of CCP's collective leadership.

Here we have to be clear that China's current political system is CCP Dynasty with a core like an emperor. The core now is Jiang Zemin and Xi Jinping has been selected by Jiang as his successor. You may find detailed description in my book *Tiananmen's Tremendous Achievements Expanded 2nd Edition*.

How much Chinese military is actually spending? I can safely say, it spends more than U.S. military as proved by the large number of expensive projects it is carrying out.

As mentioned above, China is implementing an expensive lunar program, from which the PLA (Chinese military) has drawn the technology of its anti-satellite (ASAT) technology, including hitting a satellite, rendezvousing with a satellite to blind it with spray, capturing a satellite by the robotic arm of China's ASAT satellite and destroying the internal chip of a satellite with electromagnetic pulse weapon, its ASAT defense capability (the ASAT quick response capability), aerospaceplane such as Shenlong drone and J-28 space-air fighter, HGV (hypersonic glide vehicle) and anti-ICBM missile.

For its air force, it is developing quite a few stealth aircrafts including J-20, J-31 stealth fighter jets, J-18 VTOL stealth fighter jet, the most advanced AEW&C able to detect stealth fighters and long-range strategic stealth bomber. It is producing lots of advanced military aircrafts

On the research and development project to develop aircraft engines alone, it has allocated USD16 billion funds.

For its navy, it is producing the most advanced nuclear and conventional submarines, lots of advanced 052D Aegis destroyers and 056 frigates and huge 081 and 071 landing platform docks, which may serve as aircraft carriers if China has successfully developed its J-18 VTOL stealth fighter jets.

It has imported four Zubr-class world largest air-cushioned landing crafts from Ukraine, two assembled in Ukraine and the other two in China so that China may get the technology of their production.

In addition, China is developing its Type 055 large destroyer with displacement exceeding 10,000 tons.

Chinese military has all the projects U.S. military has except the most advanced Ford-class aircraft carrier. However U.S. Air-Sea Battle by its navy equipped with aircraft carriers was what it was engaged when fighting Japan in the 1940s. For Xi Jinping's strategy for acquiring integrated space and air capabilities, large aerospace attack aircraft is the top priority.

China will spend more in acquiring such capabilities than building the most advanced aircraft carriers as integrated space and air capabilities may make aircraft carriers obsolete.

All the above is what we know from public sources. There are many secret projects that outsiders have no source of information. Based on Chinese military's activities revealed from the information in the public domain, we see that Chinese military is now spending more than U.S. military. Popularity of Xi Jinping's Chinese dream including the dream to make China militarily powerful has given rise to Chinese people's keen interest in weapon development. That certainly affects the work enthusiasm among Chinese weapon development workers.

Chinese People's Arms Craze in Imitating "Aircraft Carrier Style"

China's official TV station CCTV's footage of China's first successful landing of a jet fighter aboard its aircraft carrier has an image of two crew members on the ship wearing bright yellow vests and helmets and using their outstretched right arms to point and direct a J-15 jet, while their left arms remain behind their backs.

The image had soon gone viral. Lots of Chinese web users were so happy at the news that they posted photos of themselves imitating the crew's poses and dubbed the poses as "Aircraft Carrier Style".

Since Xi Jinping has been so successful in arousing Chinese people enthusiasm for the dream for a militarily powerful China, will he refuse Chinese military's reasonable demand for funding development of advanced weapons when the amount required exceeds by far China's military budget. He will not refuse as long as China can afford it.

Intensive Arms Race Causes Stress on Elite Engineers

China's unilateral arms race with the U.S. has put great stress on China's elite weapon development scientists and engineers who worked with dedication due to their patriotism.

The death of Luo Yang, the responsible person for the J-15 fighter project, on November 22, 2013 is a case in point.

Luo, 51, died of heart attack while overseeing the tests of J-15 taking-off and landing on the aircraft carrier.

He was an excellent engineer in China's Shenyang Aircraft Corp. that has developed most of China's jet fighters. A graduate with a master degree from Beijing University of Aeronautics and Astronautics, Luo has been awarded many times for his prominent contributions to China's weapon development. He had been paid special allowance by the state since 1999 and was appointed top management posts in the corporation, being its chairman of board and general manager at the time when he died.

From 1966 to 1976 when Luo was 5 to 15 years old, school education was almost none-existent in China due to the Cultural Revolution. However, Luo was still able to become a top engineer. It was certainly because of good family education as at that time educated parents tried their best to teach their children.

Some people regard the generation of Chinese people from Luo's age to Xi Jinping and Li Keqiang's age as a lost generation as they lost some years of schooling when they were young. I described in my book *Tiananmen's Tremendous Achievements Expanded 2nd Edition* that due to family education and the much time they had at that time, some of them, though a small proportion, became well educated through self-study under the guidance of their relatives and friends.

I regard them in my book as the youngest of the new generation of talented scholars with moral integrity emerged during the Cultural Revolution. As those scholars are in power now and will remain in power for one or two decades, I said in the book that I was quite sure that China might catch up with or even surpass the U.S. in one or two decades.

That worries me most especially because there are clear signs that China is quickly catching up with the U.S. in its weapon development.

There is unfortunately the probability under China's current political system that a despot like Mao may rise to the top. When such a despot does come to power, there is no mechanism in China to restrict or remove him. When China is stronger than the U.S., the despot may want to use China's military strength for aggression.

World people must be on their alert at that probability. That is why I have written so much about China's weapon development in my blog and in this book.

Twice Mao brought the world to the verge of nuclear war. Prominent U.S. diplomat Henry Kissinger describes it in his book "On China", but still praises Mao. In spite of the problem in China' political system, Kissinger tries to make people believe at the end of his book that people need not compare China's rise to Germany's rise before World War I. He says there will not be confrontation between China and the U.S. because the leaders in both countries may act soberly.

Kissinger even says in his book that if before World War I, European leaders had known the disaster caused by the war, there would not have been the war due to Germany's rise.

Unfortunately, at that time European leaders were not wise or sober. After World War I. Hitler was even mad in starting World War II. Mao was mad to bring the world to the verge of nuclear war. Mao did not care if half of Chinese population perished in the war.

We Chinese have to make great efforts to improve China's political system to prevent the emergence of another Mao and to establish the mechanism to restrict or remove a mad leader. On the other hand, world people certainly have to be on their guard.

You may wonder why I write so much at the age exceeding 70. I have to as I have not much time left. I have to make more people pay attention to my warning

Death of Another Younger Top Fighter Jet Engineer Less than a Month Later

On December 22, another top fighter jet engineer Guo Changchun died of heart attack at the age of 46. He was in charge of the PLA's most advanced fighter jet J-31 project. Guo was the second top engineer at the AVIC-affiliated Shenyang Aircraft Corporation to have suffered a fatal heart attack recently less than one month after Luo Yang's death. Obviously Chinese aircraft research workers are overworked.

Why Will China Surpass the U.S. Militarily?

In a previous section I described China's potential to surpass the U.S. militarily due to four reasons. Let us now make comparison between China and the U.S. First, China has a leader who knows well not only military strategy but also the technology required in modernizing Chinese military. U.S. President Obama, however, is not aware that China is vigorously conducting an arms race with the

U.S. and may soon surpass the U.S. He is losing world leadership, but still boasting that the U.S. will remain world leader for 100 years to come.

He is still boasting that the U.S. is the strongest militarily and economically but fails to see that according to his theory that Chinese currency is gravely undervalued, U.S. GDP is not so much greater than Chinese GDP in real term even if he refuses to accept the view that China will surpass the U.S. in terms of purchasing power by the end of this year.

Militarily, China has already surpassed the U.S. in some key areas such as hypersonic glide vehicle (HGV), anti-satellite (ASAT) and anti-ASAT capabilities, satellite navigation system in the area near China for accurate positioning, electromagnetic gun, etc. and will soon surpass the U.S. in aerospace plane, stealth fighter jet, etc. U.S. leader does not have the vision and wisdom to have made any corresponding efforts to maintain U.S. superiority.

Second, Chinese dream Includes the Dream for a Militarily Powerful China while American dream concerns more about people's daily life.

Third, China has abundant funds to support its unlimited budget for military development while the U.S. is hard up now.

China's tax income grows rapidly along with its economic growth. In addition, there is the income from state-owned enterprises at Chinese government's disposal. Moreover, lots of corrupt officials' assets have been confiscated in the anti-corruption campaign while the mass line campaign has greatly reduced government spending. The funds allocated to the military rise as a result of the increase in income and decrease in spending.

Obviously the increase in funds and scientists' and engineers' enthusiasm in pursuing their Chinese dream will enable China to surpass the United States as soon as possible.

Third, China can get weapon technology from Europe not only from Russia and but also from U.S. allies in the EU, especially Germany, France, the UK and Italy.

It is not merely because of Chinese willingness to pay high prices for weapons and military technology. U.S. failure to trust its allies is making them less concerned about U.S. interests. The lots of EU sales of weapons and military technology will be described in Chapter 10.

Fourth, China is able to give full play to its scientists and engineers' talents and diligence.

The American Dream

When I met my friends who have immigrated to the United States and are having successful careers, they talked much about their houses and cars and told me having ones own house and car is American people's American dream. That does not seem lofty enough.

Americans' forefathers who longed for freedom were lucky to have a place to immigrate and set up a great nation there with their lofty ideal: All men shall be equal and endowed with certain inalienable rights, including life, liberty and the pursuit of happiness.

Chinese people's forefathers certainly did not have such lofty or such immediately personal dreams. When Chinese people's forefathers had moved to California to do the backbreaking toil in building American railways, they were satisfied with a wage that was only a fraction of the wage Americans' forefathers received.

They made contributions in building up a great nation, but what did they get in return? They were discriminated against and humiliated. Local American workers protested and wanted to deprive them of the jobs that they relied on for their miserable survival. In fact, due to their country's weakness and poverty, they suffered severe discrimination and humiliation in the U.S., much worse than black people and much worse than that other Overseas Chinese suffered in anywhere else at that time.

The worst humiliation they suffered was in the 1900s. At that time, all people could enter the United States from Canada freely except Chinese people, who had to be examined nude in a cage for cattle.

What dream could those poor Chinese forefathers have then? The American Dream? They could not even be nationalized to become American citizens. Even if they might have become American citizens, they would have ranked even below black people.

Chinese Forefathers' Chinese Dream

Gradually, a Chinese dream emerged in the minds of those Chinese underdogs: They should obtain renaissance of the Chinese nation. In spite of their poverty, their donation constituted a major source of fund for Sun Yet-sen's democratic revolution as seeing the prosperity brought about by American political system, they believed that democracy might brought about China's renaissance so that when China regained its glory, they might go back to their home, sweet home to be free from discrimination and humiliation.

However, the revolution failed to bring about China's renaissance. On the contrary, the situation went from bad to worse. Later when two of my cousins graduated from high school, they could not find jobs though at that time, the number of high school graduates was quite small in China. Like those poor forefathers of Chinese people who left China for survival in the United States, my cousins had to leave home, sweet home to take jobs in Indonesia.

Later, after eight years of cruel Japanese occupation, almost all Chinese people who had suffered persecution and humiliation from Japanese invaders had the same Chinese dream as that of the Chinese underdogs in America.

What about my cousins who had become quite rich in Indonesia?

Perhaps, the environment for overseas Chinese was better in Southeast Asia. However, my cousins remained patriotic. During the war of resistance against Japan, like other overseas Chinese, they donated funds to help the Chinese government and troops. Overseas Chinese's donation was one of Chinese government's major sources of fund during the war.

My cousins too had the Chinese dream for China's renaissance. After the war, they returned to China and brought with them their wealth. They said they wanted to use their business expertise to make China rich and strong. In spite of their wealth, they were discriminated against abroad and wanted China to be rich and strong.

People outside China perhaps find the Chinese dream not lofty enough, not wise enough, not personal enough or not desirable enough, but it is the Chinese dream arisen from China's miserable past and a dream that Chinese people readily share. That is why party boss Xi Jinping wants to exploit it as a rally call.

American people must be clear about the humiliation Chinese people suffered in the U.S. Otherwise, U.S. Senate would not have apologized for the Chinese Exclusion Act in 2011.

However, they certainly do not really understand the stimulation that Chinese dream may give rise to now and the trouble it may give rise to both countries in the future.

Xi Jinping's Chinese Dream Now

In 2009, under the influence of Maoist sinocentric cosmology, PLA senior colonel Liu Mingfu published his leftist book "Chinese dream: Great Power Thinking and Strategic Posture in the Post American Era" to reject Hu Jintao's idea of China's peaceful rise and advocate instead China's "military rise". Liu believes that China's goal shall be to surpass the United States and become world number one militarily. The book was an instant success and soon sold out. However, Hu Jintao banned reprinting of the book due to its leftist theory that pursues military hegemony.

Xi Jinping, however, thinks that he can exploit it to greatly facilitate his reform. He expands Liu Mingfu's Chinese dream into a dream for the great rejuvenation of the Chinese nation.

Soon after he came to power, he brought all the Politburo Standing Committee (PSC) members to visit "The Road Toward Renewal" exhibition in Beijing. There, he said that the realization of the great rejuvenation of the Chinese nation was the greatest Chinese dream for the Chinese nation now and called on people to strive to realize the dream.

What does Chinese dream means for Chinese military?

The day after November 29 when Xi Jinping talked about Chinese dream, for the first time, Liu Mingfu received a phone call from his publisher that reprinting of his book was allowed. It seems Liu's idea on surpassing the United States and becoming world number one militarily is acceptable.

To make it clearer, when Xi Jiping visited Chinese navy on April 11, 2013, he talked about the dream for a militarily powerful China to emphasize that his Chinese dream includes the dream for a militarily powerful China.

For Xi Jinping and most Chinese people, Chinese dream means making China powerful to avoid a repetition of China's history of being bullied by foreign powers for a century. Chinese people regard U.S. support for Japan, the Philippines and Vietnam in territorial disputes as bullying China. That is why common Chinese people have such enthusiasm in China's weapon development.

Chinese TV stations have begun to broadcast programs on China's weapon development and compared Chinese weapons with foreign ones. They also have the programs on military situation. Those programs have soon become very popular.

The Chinese Dream that Defeated the U.S. in Korean War

American people are clear about the humiliation Chinese people suffered and U.S. Senate did apologized for the Chinese Exclusion Act in 2011. However, they certainly do not really understand the Chinese dream now. That may give rise to trouble to both countries.

In early 1950s, the failure to understand Chinese people's Chinese dream caused the U.S. to be defeated by the Chinese army with much inferior weapons and technology.

Some may argue that as neither side was able to drive the other away from the Korean Peninsula, the war was a draw. However, in a draw between the troops of a very weak and poor country and the extremely strong troops of 39 countries, people usually think that the weak and poor one is the winner.

Moreover, before Chinese troops entered North Korea, American troops had advanced near the border between China and North Korea. Chinese troops had driven American troops back to the border between North and South Koreas and American troops failed to regain the ground they got before the entry of Chinese troops.

Why was America defeated?

Because of the Chinese dream.

At the ceremony of the establishment of the People's Republic of China, Mao Zedong made his famous announcement "Chinese people have stood up."

What did Mao mean? He meant that since the CCP has seized state power, the Chinese people have realized and will further realize the Chinese dream of China's renaissance; therefore, the Chinese people had stood up and would no longer be bullied.

In fact, Xi Jinping is not the first to rally Chinese people by calling them to realize the Chinese dream. Since the establishment of the CCP, the CCP had tried to rally Chinese people around it, claiming that since the previous reform and democratic revolution all failed to achieve China's renaissance, only the CCP's ideology, the Marxism-Leninism, would enable China to achieve its renaissance.

As a result, most Chinese people were excited at Mao's announcement, but American politicians, at least American general McArthur did not know what Mao meant and what Chinese dream the Chinese people shared with Mao at that time. Since the Chinese people had stood up and would no longer be bullied, how could they fail to take action when foreign troops came near Chinese border?

If McArthur had understood what Mao meant and the Chinese dream, he would not have regarded Chinese government's warning as bluffing when his troops went near Chinese border. Nor would he have believed that China would not have sent its troops into North Korea when he did not find any Chinese troops during his personal reconnoiter flight above the area where Chinese troops were hiding.

In fact, if McArthur had understood the Chinese dream, he would have sent airplanes to reconnoiter at night with illuminating projectiles. He would have found tens of thousand Chinese troops were going to North Korea through six temporary bridges.

American people would not have had the erroneous idea that Chinese soldiers had been brainwashed when they were amazed at Chinese soldiers' bravery that was inspired by the Chinese dream.

However, American defeat in Korean War was not only a disaster for General McArthur, but also an even greater disaster for the Chinese people.

The victory made Chinese proud and quite a few Chinese who had completed their further study abroad came back to help the CCP build up China with the mentality mentioned in my book.

However, the victory enabled Mao to establish his personality cult.

Moreover, Mao and quite a few Chinese communists were carried away by the victory and thought that since they could defeat the most advanced troops in the world, they could achieve similar quick victory in their economic development and catch up with the United States in a short period of time.

Their arrogance resulted in the absurd Great Leap Forward and then the worst man-made famine in Chinese history.

However, that was not the end of China's misery. The power struggle caused by the economic failure gave rise to the Cultural Revolution that brought greater disasters to Chinese people especially their intellectuals.

Americans' Ignorance of Xi Jinping's Chinese dream

Now, Xi Jinping, the newly selected party leader, talked about the Chinese dream again. Those who know nothing about the Chinese dream cherished by Chinese people for several generations regard the Chinese dream talked about by Xi as something new. In a six-page long article on one of the two major American weeklies, the author says, "Xi has exhorted citizens to pursue 'national rejuvenation' and a 'Great Chinese Dream'" as if national rejuvenation and Chinese dream are two separate things. As a result, the author asks, "What exactly is the Chinese dream?"

Xi clearly said in his speech, "In my view, to realize the great renewal of the Chinese nation is the greatest dream for the Chinese nation in modern history."

In the article, the author says, "The symbolism is potent but vague on details." He forgets the context of Xi's speech. Xi first said in his speech, "lagging behind leaves one vulnerable to attacks and only development makes a nation strong."

The goal of revival is first of all to make China so strong that it will no longer be bullied.

Since the U.S. announced its return to Asia and rallied around it those who have maritime territorial disputes with China, China has begun its arms race with the U.S. In 2012 alone, China established its Beidou Navigation System that covers the Asian-Pacific area, commissioned its first aircraft carrier, developed its first carrier-based fighter J-15, conducted successful test flights of its stealth fighter J-31, VTOL stealth fighter J-18 and a new fighter jet J-16 about which no information has been revealed, commissioned its second and third 20,000-ton landing platform docks and launched two 052D Aegis destroyers.

According to Russian media, China sees itself in the remote future as a military superpower equal to and even in some areas surpassing the United States.

Chinese media hold that the U.S. relies on its advanced air force in war. If China surpasses the U.S. in its air force, the U.S. will have nothing to rely on in bullying China.

That seems China's way of "subduing the enemy without fighting", a strategy advocated by Sun Tzu.

The surge of patriotism due to maritime territorial disputes has caused China's aircraft research workers to work so hard that China is developing its advanced aircrafts and other weapons at an astonishing speed.

Xi's rally call, the Chinese dream, will certainly add stimulus to their enthusiasm and further speed up China's weapon development.

Chinese Dream Helps Xi Jinping Build Up His Powerbase

Achievements in Weapon development are certainly important now for China in dealing with the territorial disputes, but for Xi, the rally call plays a vital role in consolidating his power.

Being a new leader Xi needs to establish his powerbase. The need is much more urgent as Xi has two very tricky tasks: eliminating corruption and deepening the economic reform. Like his predecessor Hu Jintao, Xi encounters the seemingly insurmountable resistance from vested interests and the conservatives.

Realization of the Chinese Dream will enable Xi to take harsh measures in fulfilling the tasks as corruption robs the state of its resources for weapon development while further reform will enable the state to achieve faster growth so that it will have more income from taxation and state-owned enterprises for weapon development.

Like the Korean War for Mao, the victory in a war against Japan and even against the U.S. will enable Xi to firmly establish himself as a leader with absolute power; therefore, Xi has to succeed in making Chinese air force as strong as the U.S.'s.

That is why the *Liaoning* aircraft carrier, J-15 carrier-based and J-18 and J-31 stealth fighter jets, etc. are major parts of the Chinese dream.

A Loftier Chinese Dream

I am worried that if U.S. support for the countries in their maritime territorial dispute with China results in a war between China and the U.S. and if China wins, Chinese leader will enjoy very high prestige which perhaps may give rise to some personality cult that may bring disasters to China as Mao's personality cult did in the past.

Therefore, I hope while realizing the age-old Chinese dream, Chinese people begin to cherish an even loftier Chinese dream: the implementation of Chinese constitution. However, only a relatively small proportion of people share that lofty Chinese dream; while the Chinese dream Xi advocates is very popular now. It proves how skilled Xi is in exploiting popular feeling.

Compared with Xi's Chinese dream, the even loftier Chinese dream for the implementation of Chinese constitution is much less popular for the time being.

It took more than one decade for China's democracy movement to grow from the Democracy Wall in 1978 into a national campaign in 1989, the Tiananmen Protests, which were unfortunately suppressed by the military.

The Chinese dream for the implementation of Chinese constitution will also grow if people fight for it peacefully without violence. When it grows into a popular national campaign perhaps two decades later, it would not be suppressed by the military as due to the realization of Xi's Chinese dream, the military will have grown into a well-educated professional army unwilling to point their guns at innocent people.

We can already see such signs now. The most prominent sign was a commentary published on the sensitive date of June 5 on the party's mouthpiece People's Daily by Zhang Yang, the political commissar of People's Liberation Army's Guangzhou military command, in which Zhang warns against abusing the deployment of troops for non-combat missions.

It indicates the military's reluctance to be used to suppress people. That was why Deng said after Tiananmen Massacre that China was "lucky" then because elders like him were still alive, hinting that if all the elders had died, no one would have been able to send troops to suppress the students.

7. Integrated Space & Air Capabilities

In Chapter 5, I said according to a report that the U.S. has most recently issued, China is testing Shenlong aerospace fighter that is able to reach the U.S. in one hour. I also mentioned J-28 fifth-generation stealth fighter China has been developing.

The above information is not official and may be mere speculation as China always keeps confidential its development of advanced weapons. However, as Xi Jinping stressed development of integrated space and air capabilities during his recent visit to PLA air force, we shall believe that news in this respect, even though speculation, deserves attention. However, most of the information below has come from either Chinese official sources or reliable media.

Concerning China's integrated space and air capabilities, we will first give information on its aerospaceplanes as large aerospace bombers are most powerful. Certainly China's ICBM interception and ASAT capabilities are also important. The latter has been quite intensively covered above; here we will pay greater attention to the former.

China's Mystic Space Warfare Force

The information below comes from authority sources.

On April 1, 2014, Zhuang Fenggan, China's aeronautics authority, revealed to qianzhan.com reporter China's constant research and development of aerospaceplanes for a long time. Xian Luliang Strength Test Center of China Aviation Industry Corporation 1 was in charge of testing aerospaceplanes in the period of China's eleventh five-year plan (2006-2010). If so, China has conducted test of aerospaceplane since 2010 the latest. Zhuang believed that aerospaceplane was the most important platform to be developed for aerospace weapons. Since then, China's mystic space warfare has gradually been known to the public.

Aerospaceplane or aerospace aircraft is a reusable flying vehicle capable of traveling between atmosphere and outer space with advanced technology for both military and civil purposes. Zhuang stressed that it could be used as a platform for weapons, but China would by no means change its policy of peaceful utilization of outer space. According to him, the development of aerospaceplanes concerns China's space and air security. An aerospaceplane is characterized by high speed, maneuverability and undetectability and long range. Mr. Zhuang believed that compared with existing aircrafts in the world China's aerospaceplane will be one of the best in terms of shape, functions and performance.

China began fundamental research for aerospaceplanes at the beginning of the 21st century. Over the past few years, China has invested lots of financial and human resources in such research and development, but he did not reveal the timetable for the emergence of China's aerospaceplane.

In addition, official media have reported that in the period of the eleventh five-year plan, China Aviation Industry Corporation 1 undertook the research, manufacture and tackling of key problems for major flying vehicles. Luliang Strength Test Center in the Corporation undertook major testing tasks for aerospaceplanes. The Chinese Academy of Sciences also commenced the project of international cooperation of "the team of high-temperature gas flow research for aerospaceplanes"

In fact, when Chang'e I Lunar Exploration Satellite was launched, some media hinted that the PLA was busy making preparations for the establishment of China's space warfare force. Only it takes time to develop a reusable flying vehicle for traveling to and fro between earth and space. However, as Zhuang had revealed China's plan for the development of its aerospaceplanes, we can say that the decision has already been made for the PLA to make preparations for the establishment of a space warfare force.

It is said that a space station is ideal for watching the earth; therefore, it can be used as space command center and a base to test the deployment and use of space weapons. U.S. refusal of China's repeated request to join the international space station make it justified for China to build its own permanent space station.

Now, the U.S., Russia, India, Japan and even South Korea are setting up their space warfare forces. It is said that China began setting up its own space warfare force in 2000 mainly in response to U.S. decision on development of missile defense system referred to as "the New Space War". China's goal was not to lag behind in the competition for control of space that would soon begin.

Qianzhan.com's report described China's ASAT capabilities mentioned earlier in this book. In addition to what I have said above, it says that China is advanced in high-energy laser technology and has the capability of shooting laser beam from its space base to blind enemy low-orbit satellites.

Graphene—China's Super New Material for Spacecrafts and Aircrafts

Graphene is a material that most people are not familiar with, but researchers in many countries are busy working on it. It is a new material formally discovered in 2004, for which two British scientists Andre Geim and Konstantin Novoselov became Nobel Physics Laureates in 2010. Its special characteristics make it possible for the production of transparent mobile phone as thin as a piece of paper, batteries able to be fully recharged in a minute, bullet-proof shirt and other science-fiction products.

For real military application, graphene is very thin and nearly transparent pure carbon one atom thick. It is remarkably strong for its very low weight, more than 100 times stronger than steel. This makes it the best materials for aircrafts and spacecrafts.

However, there is drawback: It is forbiddingly expensive, 15 times more expensive than gold. That may be a problem for the United States as it has to reduce military spending due to its huge budget deficits, but no problem for China that has an unlimited military budget.

China certainly would not grudge however large spending for this revolutionary new material in order to obtain integrated space and air capabilities.

Sources related to the Industrial Aviation Material Research Institute of China Aviation Industry Corporation revealed to China's Global Times on May 27, 2014 that they have obtained the technology for mass production of graphene film and powder and successful used graphene to make an alloy of graphene and aluminum, the first such alloy in the world, with exceptional qualities for aircrafts and spacecrafts.

In Chapter 6 the section with the subhead "Chinese Military's Unlimited Budget", I said "China has abundant funds for military development and mentioned its growing tax and SOE's income, income from confiscating corrupt officials' assets and savings from reduction of government spending.

Let me give you an example. Before the mass line campaign, a quite rich province spent three times its budget in building luxurious government offices and official housing and providing officials with luxurious cars, banquets and travels abroad. Since the campaign, such luxuries have been strictly forbidden and the province has to reduce spending within its budget. Lots of funds thus saved can be used for military spending.

There is still the question why shall China spend such a lot on its weapons.

Because it is busy conducting an arms race with the United States in order to surpass the U.S. in military strength.

I described the beginning of arms race in the section with the subhead "Upset by U.S. Containment, China Began Its Arms Race with the U.S." in Chapter 6.

People outside China know Chinese leader Xi Jinping's Chinese dream for the rejuvenation of the Chinese nation, but perhaps are not aware of the connection between Chinese dream and the arms race. Xi's Chinese dream is based on China's history of being bullied by foreign powers for a century. He believes, "lagging behind leaves one vulnerable to attacks and only development makes a nation strong."

Therefore, as mentioned above he borrowed the Chinese dream from Liu Mingfu and expanded it into a dream for the rejuvenation of China, but concerning the military, in the face of U.S. military threat, Xi keeps Liu's idea to include making China militarily powerful into Xi's Chinese dream.

Readers shall be clear that if the U.S. persists in interfering with China's claim to the area within its nine-dash line, a war between China and the U.S. is perhaps unavoidable.

That is due to the clash of civilizations between China and the U.S. as described in Samuel Huntington's gifted book *The Clash of Civilizations and the Remaking of World Order*.

China regards the area in the nine-dash line as the precious legacy left behind by its ancestors. Its ancestors had claimed the area for centuries. That is legitimate according to thousands years of Chinese civilization but unlawful according to U.S. civilization that has a history of only a little more than two centuries. The U.S. regards the international law adopted by the UN much later as legal basis.

Can China allow the territorial waters it has had for centuries to be deprived by a superpower? Certainly not. "Lagging behind leaves one vulnerable to attacks and only development makes a nation strong." In Chinese view, China is now developing the capabilities to resist the bully from a superpower.

In fact, the U.S. has enough financial resources if it conducts arms race with China wisely, but it wastes its precious resources in its outdated strategy of Air-Sea Battle.

With exceptionally good materials such as graphene, China will develop a hypersonic aerospace bomber carrying more than 100 hypersonic missiles made of similar good materials to destroy U.S. Air-Sea capabilities in minutes.

Why can an aerospace bomber carry more than 100 hypersonic missiles? Because the missiles on the bomber are flying at the hypersonic speed of Mach 22 along with the bomber. They need no fuel or engine to attain their speed but the little fuel to control them to hit their targets. There is no defense against such hypersonic missiles in the near future.

If the U.S. does not change course, it is doomed to failure in future war with China.

China Tests Mach 10 Hypersonic Weapon

The Washington Free Beacon said in its report on January 13, 2014 that according to an anonymous Pentagon official, China conducted test of its hypersonic glide vehicle (HGV) on January 9, 2014. Pentagon gave the HGV the codename WU-14 and regarded it as a hypersonic weapon.

As such a HGV can penetrate anti-ballistic missile system, China's test of its HGV was a great step forward in its development of integrated space and air capabilities with new-type strategic nuclear and conventional missiles.

It is said that the HGV is installed in the transformed carrier rocket of an ICBM. It is launched from the ground, separates from its booster and then flies at high speed without driving force. Its speed may reach Mach 10.

The U.S. tested its HTV-2 HGV in 2010 and 2011 but failed. Russia carried out such tests too in the period from 2005 to 2009, but no information about the tests is available.

U.S. military spokesman Jeffrey Pool said that the U.S. had noticed the Chinese test but gave no comment in order to avoid misjudgment.

China's HGV Is an "Extensive Threat" to America—U.S. Media

U.S. *Aviation Technology and Space Weekly* carries an article in its January 27, 2014 issue titled "U.S. navy regards China's hypersonic guide vehicle as the part of Chinese weapons with extensive threat", pointing out according to U.S. navy, the Mach 10 hypersonic guide vehicle (HGV) China tested on January 9, 2014 reflects China's foresight on future war. Once China is able to apply that technology, it will have a weapon that can challenge all existing missile-defense systems and widen the range of its ballistic missiles. It takes a few years for such weapons to be usable depending on solution of the difficult issues of controlling its guidance and making it hit accurately.

The report says that the HGV test enables great improvement in China's anti-ship ballistic missile (ASBM) and may probably bring about China's second generation of ASBM. According to U.S. China military expert Richard Fisher, the DF-26 missile mentioned in rumor perhaps has a HGV warhead and will have a longer range of 3,000 km than DF-21's 2,000 km. If China's DF-31 ICBM is installed with an HGV warhead, its range will be lengthened from 8,000 km to

12,000 km.

The report believes that the U.S. has to develop directed energy weapon to counter such missiles and the existing missile-defense systems cannot intercept missiles with a speed exceeding Mach 5.

In some analysts' opinion, China's test proves that it is highly necessary for the U.S. to develop directed-energy weapons as only such weapons are able to hit unexpected targets quicker than Mach 5. The U.S. is developing such weapons but has no idea when such weapons are required and may be used.

According to the report, it takes more time to detect an HGV than a ballistic missile as it may conduct the maneuver of pulling up when it enters the atmosphere so as to fly along a relatively flat and straight trajectory towards its target. As a result, there is less time to respond or intercept. As an HGV can make some aerodynamic maneuver, an interception missile must have better maneuverability to hit it. Moreover, as HGV technology has lengthened the range of a missile, the midcourse of the missile that is relatively easy to intercept may be further away from its target and may therefore be beyond the range of defense missiles.

Better HGV Makes China Leader of Weapon Development

In the past, the U.S. was the leader of weapon development, while China had to make great efforts to develop asymmetrical advantage to counter U.S. military superiority. However, due to lack of funds and technology, such asymmetrical weapons merely managed to resist U.S. superiority, but could by no means be regarded as superior to U.S. ones to constitute a threat to the U.S.

Now it seems to be America's turn to develop weapons to counter China's growing superiority.

A clear sign of that is China's successful test of a Mach 10 hypersonic glide vehicle (HGV) in January 2014. Unlike its usual practice in the past, Chinese Defense Ministry did not remain silent to keep it confidential but responded on January 15 to U.S. media's sensational reports on the threat such weapon constitutes to the U.S. The ministry pointed out that it was normal that China conducted the scientific test within its own territories according to its own plan and that the test was not directed at any country or specific target.

Pentagon has been worried by the test and urgently wanted to catch up. It admits that China's HGV is advanced. It is aware of Russian efforts in developing HGV but believes Russia lags behind China.

U.S. navy believes that China develops HGV for its anti-ship missiles as even the best anti-missile system provides no defense against hypersonic missiles. However, the threat of HGV to warship is nothing compared with ICBM or SLBM equipped with HGV nuclear warheads.

Now, the Chinese HGV has the speed of Mach 10, but the third test of American HGV X-51 last year, though successful, only achieved a speed quicker than Mach 5.

There is the worry that if China's DF-26 aircraft killer missile uses HGV warhead, U.S. aircraft carriers will be in trouble within the range of such missile.

Previously, high Pentagon officials said that in spite of the reduction in U.S. military budget, U.S. development of HGV has not been affected, but now they have asked Congress for more funds.

Sharing the worry, U.S. congress allocated $70.5 million for development of HGV next year. It proves that the key issue remains that U.S. military has limited budget while its Chinese counterpart has unlimited budget.

HGV is top priority as U.S. experts believe mastery of the weapon will enable a country to dominate the world. Moreover, China is determined to surpass the U.S. in its arms race with the U.S. to resist U.S. bully over the South China Sea issue. Chinese President Xi Jinping said, "lagging behind leaves one vulnerable to attacks and only development makes a nation strong." He wants Chinese people to share his Chinese dream including the dream to make China militarily powerful.

To plant his Chinese dream deeper in Chinese people's minds, Chinese government spent six months to make a five-episode TV series on Chinese dream and finished showing one episode a day as Chinese Official TV's prime time program before the end of May, 2014.

Xi Jinping wants to use his Chinese dream to overcome corruption, officials' despotism, luxurious life style and bureaucratism and deepen China's reform. The resistance from vested interests seems insurmountable. However, like Jiang Zemin, he is helped by the United States.

In my book *Tiananmen's Tremendous Achievements Expanded 2nd Edition*, I point out that U.S. victory in Gulf War helped Jiang Zemin take control of Chinese military as the victory caused panic in Chinese military so that it provided Jiang with the opportunity to display his talents in modernizing Chinese troops and establishing his control in the course of modernization.

Now, by interfering with China's territorial disputes at the South China Sea, the United States helps Xi Jinping to rally all Chinese people including diehard conservatives around him to support his anti-corruption and mass-line campaigns and thorough reform.

Previously, the large conservative faction, though had lost its charismatic leader Bo Xilai, remained a powerful opponent to Xi Jinping especially his thorough reform. However, they had the desire to make China surpass the U.S. militarily. They supported Xi's struggles against corruption and officials' despotism and luxurious life style but strongly oppose Xi's thorough reform.

U.S. interference has infuriated Chinese people and given rise to a surge of patriotism. It has enhanced conservatives' desire to make China militarily powerful. That makes it easier for them to accept Xi Jinping's reform that aims at making China economically strong so that China will have more resources to surpass the U.S. militarily. Contrary to U.S. intention, U.S. efforts to contain China help Xi Jinping establish his position as paramount leader of both civilians and troops with undisputable authority. As a result, Xi will be able to successfully carry out his struggles and reform and make China much richer and more powerful than the U.S.

In this respect, the U.S. is its own enemy.

U.S. Adjusts Plans in Response to Chinese Weapon Development

Due to lack of funds, instead of creating its advanced weapons to force China to make efforts in respond to U.S. efforts, the U.S. has to watch what China has been developing over the past year and make corresponding amendments. The U.S. has adjusted four major plans on its equipment and technology including littoral combat ship, Virginia class nuclear submarine, DDG1000 destroyer and long-range attack bomber.

The original U.S. plan was to build 50 littoral combat ships, but has now reduced the number to 32 and its navy is discussing the type of warship to replace such warships. It wants the replacement warship to have air-defense missiles and small Aegis radar.

It previously wanted to use DDG1000 destroyer as its major warships but now only wants to build 3 of them.

Why?

DDG1000 is advanced mainly in its vertical launch system (VLS) and radar. However, when China's first Type 052D was commissioned on March 21, 2014, its more details have been revealed to show its VLS is by no means inferior to DDG1000's.

90

A launch tube in 052D's VLS can launch and control 4 different kinds of missiles, including air-defense, anti-ship, anti-submarine and ship-to-ground missiles of different sizes. The maximum diameter of the missile is 850 mm, bigger than DDG1000's MK41 VLS.

Unlike U.S. VLS that is only capable of hot launch or Russian VLS, cold launch, 052D's VLS tubes are capable of both cold and hot launches of missiles and they also have CCL devices so that the smoke in hot launching of a missile is discharged through a concentric tube.

There are separate launch control electronic equipment for the control of each missile launched to simplify the procedures of control so that there are no longer the complicated control procedures of multi-layer transmission of information and orders.

The new type active phased array radar used on 052D now has a liquid cooling system instead of the air cooling system in Type 346 phased array radar on 052C that has a curved case to increase the area of contact for cooling. In the new radar, liquid goes through the antenna for cooling. As a liquid cooling system has bigger cooling capacity and the contact area is bigger in the antenna array, the new radar must have greater power and better-sustained performance.

Due to the above-mentioned advanced equipment and functions, a DDG1000 is better than 052D only in bigger displacement and 16 more launch tubes. However, China is designing 12,000-ton Type 055 destroyer bigger than DDG1000. When 055 is commissioned, DDG1000 will be inferior. Therefore the U.S. has to scrap its DDG1000 plan and suddenly decides to spend $18 billion to produce 10 Virginia class submarines in haste to deal with Chinese advanced destroyers.

That clearly proved that when the U.S. formulated their plans for DDG1000 and littoral combat ships, it underestimated China's ability to develop warships more advanced than those the U.S. planned as its newest best weapons.

In the past, the U.S. always believed that it always had the newest ideas in weapon development, but now the U.S. has lost its leadership in weapon development.

China's First Vibration Test in Hypersonic Wind Tunnel, Vital for Hypersonic Weapons

On March 19, 2014, xinhuanet reporters learnt from the State Administration of Science, Technology and Industry for National Defense that China has for the first time successfully carried out vibration test in a hypersonic wind tunnel.

The test was designed by China Aerospace Science and Technology Corporation (CASC). It is a very important test for forecasting the pneumatic elasticity of the hypersonic glide vehicle (HGV) China is developing.

According to experts, as an HGV is affected by the characteristics of the hypersonic airflow field and pneumatic heating and control, pneumatic elasticity is a very complicated issue. Pneumatic elasticity refers to the structural deformation of a flying vehicle caused by increase in speed. For example, the quicker an aircraft, the greater the deformation the high-frequency vibration may cause to the wings and other components of the aircraft.

The successful development of the hypersonic wind tunnel will greatly facilitate China's research and development for hypersonic weapons.

Large Transport Spaceship Being Developed for Aerospace Bomber

I mentioned in Chapter 5, according to Russian media's speculation, due to the breakthrough in the technology to reduce the size of high-temperature air-cooled nuclear reactor, China's new strategic bomber will be nuclear powered.

The speculation was based on the information that China wanted the bomber to stay in the air for three months. However, I said that if the bomber was in a low orbit around the earth, it might remain in orbit for three months by inertia. Only the orbit will be lowered two km a month. If the orbit is 120 km above the earth, the bomber can remain in an orbit 114 km above the earth three months later. That will be quite safe for it.

If that is the case, there is no need for a nuclear reactor to power the bomber, as the reactor is not only expensive and heavy but also difficult to develop. A conventional rocket will be able to send the bomber into its orbit. The bomber only needs the rocket engine to control its soft landing on earth. To send the bomber into its orbit, China needs a large rocket to carry a heavy load instead of the one it uses now to send its relatively small spaceships to its Tiangong 1 temporary space station. There has been report that China will test such large rocket soon.

Zhang Bainan, chief designer of manned spaceship system for China's space station, is an NPC (National People's Congress–China's parliament) deputy. On March 11, 2014, he told a Beijing Times reporter that smooth progress had been made in developing China's cargo transport spaceship that will be in charge of sending propellants, supplies for life in space and scientific research equipment and bringing back scraps and space rubbish.

The spaceship is now undergoing tests. When it has been found mature enough, it will carry out routine travel to the space.

In addition, Mr. Zhang told the reporter that he had submitted a proposal to NPC a motion for a law governing space activities. In the past, such activities are all carried out by state-owned military industrial enterprises and are therefore governed by the State. However, China's further reform will open military industry to the private sector. Then there shall be laws and regulations to govern commercial space activities.

Huge Space Transport Rocket Yuanzheng I Will Serve China's Moon, Mars Projects

Later the same month, Liang Xiaohong, Party Secretary of Institute No. 1 of China Aerospace Science and Technology Corporation, said after he attended CPPCC (Chinese People's Political Consultative Conference) annual session in Beijing, that maiden flight of China's space ferry Yuanzheng I would be conducted within the year. The success of the maiden flight would be a forward stride in China's development of space technology.

The space ferry can support China's space exploration of the moon, Mars, etc. change the orbits of existing satellites to lengthen their lifespan or enable them to fulfill additional tasks, send several satellites in one travel and remove space rubbish. It involves the technology of using liquid fuel in the last stage of the rocket so that it can be reignited several times for several tasks.

As the last stage is able to work for 6.5 hours continuously, it can send several satellites far away from one another.

On the other hand, according to Chinese official newspaper Global Times, on April 2, 2014, a U.S. space news website published an article, stating that China is going to launch its largest rocket so far. The rocket codenamed Changzheng V (which must be the Yuanzheng 1 referred to by Liang Xiaohong above) will further speed up China's progress in its space project.

An American senior analyst regards China's space project as not only military but also scientific and commercial. This reform in its goal has been reflected in the establishment of a new space launching site in Hainan Province.

A professor in U.S. naval war college says that the preparations for the maiden launch of Changzheng V have been very difficult. The launch of the rocket has been delayed many times. However, "China's persistence in the project reflects its will to become a space power. For the fulfillment of China's 'one step manned space travel' plan formulated in 1993, a carrier vehicle with great boosting force is perhaps the most important hardware in the 'puzzle'."

Changzheng V rocket is China's largest rocket ever with a diameter of 5.2 meters. It has to be assembled at the launching site as a train carrying it cannot pass a railway tunnel. That means China's space program has been freed from railway restriction.

Note: All the above three reports fail to mention how heavy a load the large rocket can carry. Up to today, it has remained a mystery.

China to Be the First in Using Laser Propulsion for Its Aerospaceplane

According to speculation by British magazine *New Scientist* and Russian magazine *Newsland*, China has begun the research project of using laser propulsion for its aerospaceplane. With laser propulsion, an aerospaceplane will have very strong power to carry substantial load and fulfill various tasks. As a result, China will vigorously defeat the U.S. and Russia in the competition to develop aerospace equipment.

China's official media has published a report on the official establishment of China's state key laboratory for laser propulsion, which marks a sound step forward in China's exploration for the highly efficient aeronautical propulsion technology.

British *New Scientists* magazine says that in laser propulsion a high-energy laser is used to heat something and turn it into gas to propel the flight vehicle. It has the advantages of great propulsion force and low cost and can be used for lots of purposes such as launch of small satellite, removal of debris along the orbit of the flight vehicle and control of the status and orbit of a satellite. Using that technology the launching cost will be reduced to a few hundred dollars per kg from the nearly $10,000 per kg when rocket is used as launch vehicle now.

According to speculation by Russian *Newsland*, obviously China set up the state's key laboratory not merely for the purpose of the above-mentioned application of the technology. Very probably, China will begin to develop a new type of aerospaceplane propelled by laser.

American, Russian and other countries' scientific research proves that solar energy is not adequate to propel a large aerospaceplane. As a result, great importance is attached to the technology of laser propulsion. Using powerful laser to propel a flying vehicle sounds like a science fiction but is quite feasible.

According to speculation, China's state key laboratory for laser propulsion has begun testing a technology called "laser ablation". It uses laser to ablate a special metal at the tail of a flying vehicle. The gas resulting from evaporation may provide driving force. There may be another test to set up a huge sail with a special metal coating on an aerospaceplane. A laser from earth will "burn" the metal coating to generate a strong propelling force. In theory, both the above technologies can propel a large aerospaceplane.

8. Great Improvement in Submarine, ICBM & SLBM

Busy Tests of China's ICBMs and SLBMs

How many ICBMs, submarine-launched ballistic missiles (SLBMs) and nuclear warheads China has is China's well-guarded secrets. The estimate varies greatly. The previous estimate is China has 30 ICBMs, which is an unreasonable underestimate because China has been developing nuclear weapons for over 4 decades.

In March a certain country's strategic intelligence agency has got information that according to the new secret intelligence obtained and new data collected by spy satellites, China has as many ICBMs combat ready as its medium-range ballistic missiles, the number of which is hundreds as revealed by Chinese source.

As for nuclear warheads, Some Western analysts believe that the number of China's nuclear warheads does not exceed 250. However, Russian *Military Messenger* weekly published an article on July 24, 2013 titled "China's grand nuclear gift—China may have the largest nuclear arsenal in the world". The article says, "In the 1990s, China was able to make more than 140 nuclear warheads p.a. Even if some of the old warheads have been dismantled, the estimate of 250 warheads cannot even be regarded as an awkward joke."

The author of the article, a deputy director of Russian Institute for Political and Military Analysis, believes that from the perspective of production capacity, there are at lease a few thousand and even up to ten thousand nuclear warheads in China's nuclear arsenal.

Such information was revealed at that time due to the dire possibility of a war between China and Japan that may involve the U.S. U.S. underestimate of China's nuclear arsenal may lead to nuclear attack against China. That is why since the intensification of Sino-Japanese territorial disputes there has been more news about Chinese ICBMs and SLBMs.

On December 13, 2013, China conducted the second flight test of its newest long-range missile DF-41, a road-mobile, MIRV-capable ICBM with a maximum range of 14,000 km, enough to cover the entire United States. There has been no Chinese official information about DF-41, but according to U.S. intelligence, it is capable of carrying 3 to 6 independently targetable nuclear warheads. There has been report that DF-41 is capable of breaking through U.S. anti-missile network.

That was the second test of DF41. The first test took place on July 24, 2012.

Less than 10 days later, on December 22, 2013, China test launched a JL-2 SLBM.

China's official media huanqiu.com posted 26 photos of the above two tests. As the website belongs to Global Times, a subordinate newspaper of Chinese Communist Party's mouthpiece People's Daily, we regard the website as official.

The website's report contains 26 photos on the ICBMs, but only the second photo shows the test of JL-2 SLBM being launched from a submarine. JL-2 has a range of 8,000 km. It can only reach the east part of the United States if it is fired near Chinese coast. It is also MIRV capable and carries three independently targetable nuclear warheads.

China's Successful Test of JL-3 SLBM Able to Hit Anywhere in the U.S.

In mid February 2014, Japanese newspaper Yomiuri Shimbun disclosed that a PLA nuclear submarine successfully launched China's most powerful JL-3 SLBM and hit a target at a desert in Xinjiang.

JL-3 is China's third-generation submarine launched ICBM. It uses a Changzheng-2F carrier rocket with reduced size. The rocket has its booster removed but has been installed in addition with warheads and solid fuel. Its range is 5,000 km longer than JL-2. It carries 5 to 7 independently targetable 350kt nuclear warheads and is expected to be deployed in Type 096 nuclear submarines.

Analysts of U.S. Navy Times believe that if JL-3s are deployed in China's new generation of nuclear submarine, the entire United States will be within the range of the missiles no matter where they are. This enables China to have the capacity of multiple nuclear strikes after being hit by its enemy as long as any of its submarines have survived.

A Russian military expert believes that the launch of a JL-3 SLBM proves that China's nuclear deterrence has been upgraded from tactic to strategic level. China has integrated the technology of new submarines and new missiles to meet the requirements of actual war.

However, in the above-mentioned test the JL-3 was launched from a submarine in the Yellow Sea and hit a target in Xinjiang, western China 8,000 km away. The distance is within the range of JL-2. The test did not prove JL-3 has a range longer than JL-2.

On August 4, 2014, qianzhan.com said according to Japan's Yumiuri Shimbun. A few days ago, a nuclear submarine of Chinese navy successfully launched a new-type JL-3 submarine-launched ballistic missile (SLBM) from the Atlantic.

The missile hit its target in a desert in Xinjiang. That test has really proved that JL-3 has a much longer range than JL-2 and is able to hit anywhere in the United States.

The missile enables China to have real second-strike capability without limitation by the location of its strategic submarine. China has thus become a true nuclear superpower.

China Successfully Intercepted an ICBM

China successfully conducted its initial ground-based midcourse missile interception in January 2010.

In January 2013, China's official news agency Xinhua published a brief report on China's "ground-based midcourse missile interception test within its territory" on January 27, 2013.

This is an emerging military technology aimed at destroying ballistic missiles in midcourse of its flight after the above-mentioned test in 2010.

"The test has reached the pre-set goal," the report quoted an anonymous Defense Ministry official as saying. "The test is defensive in nature and directed to no other country."

Due to the successful tests, China became the second country after the U.S. that had successfully intercepted an ICBM. That would certainly cause concerns among its neighbors.

The official did not say whether any missile or object was destroyed in the test.

The report said that no other details had been given about the test by Chinese military, but weapon experts believed that through such a test China could develop the air defense to intercept warheads from space.

PLA officials and documents have revealed in recent years that anti-missile technology is one of the key projects in China's defense budget, which has grown by double-digits for many years.

China's ICBM Interception System

In its recent post titled "HQ-19 Anti-missile Interception Missile", U.S. global security.org website reveals its speculation based on published information on details of China's HQ-19 anti-missile interception missile. It believes that HQ-19 is similar in functions to U.S. THAAD anti-missile interception missile and that HQ-19 played a major role in China's two ground-based midcourse anti-missile tests in 2010 and 2013.

However, the speculation is proved wrong by a recent blurred photo of the HQ-19 missile test appeared at a Chinese website. The missile is more similar in functions to the lengthened-range THAAD or Standard Missile 3 that the U.S. is developing. It is capable of intercepting a target inside or outside the atmosphere.

Exposure of HQ-19 attracts people's interest in China's large, complicated and ambitious plan to develop its anti-missile arsenal.

Foreign media believes that China's development of its midcourse anti-missile technology aims mainly for national security instead of nuclear strategic balance. They believe that Chinese anti-ballistic missile system consists of 6 kinds of missiles: HQ-9B, HQ-19, HQ-26 (similar to ground-based Standard Missile 3), HQ-29 (similar to PAC-3), DN-1 and DN2 (similar to U.S. GMD). There are three layers of defense in the system. The first is midcourse interception mainly by DN series of missiles to intercept missiles outside the atmosphere. It is the key layer of China's missile defense system. The second is a layer to intercept missile inside, outside or at the edge of the atmosphere. It mainly relied on HQ-19 and HQ-26 for missile interception. The third layer is the terminal stage interception layer that mainly uses HQ-9B and HQ-29 for interception.

China succeeded in testing midcourse anti-missile interception technology in its DN-1 missile on January 11, 2010. Later, it successfully tested the technology again on January 27, 2013. Analysts believe that China has obtained initial mastery of the technology.

DN-1 and later DN-2 are China's most advanced midcourse interception missiles similar to U.S. GMD system. However, as they lack the support of effective early warning radar, they remain at testing stage.

Foreign media believes that China has obtained the technology in order to intercept the ICBMs from large nuclear powers. That is not correct. It is impossible for China to intercept such a large number of first-strike ICBMs a large nuclear power has. China's strategic nuclear deterrence relies on its second-strike capabilities. The anti-missile intercept missiles can only be used to intercept a few ICBMs fired at China by mistake.

The midcourse intercept technology is mainly used to intercept the small number of ballistic missiles from a small nuclear power such as India.

2nd Artillery Corp Expert: China Lags Much behind the U.S. in ICBM Interception

The above-mentioned China's successful ground-based midcourse missile interception test raised great concerns as it was conducted a day after the United States carried out a similar test on January 26.

Some media believed that it indicated that China's anti-ICBM technology was as good as America's.

In an interview in late January 2013, a Second Artillery Corps expert who had studied strategic missiles for 45 years, told *Southern Metropolis* that indeed, the United States and China had so far been the only two countries that had independently carried out the test of ground-based midcourse missile interception technology, but there was still quite big disparity between the two countries. China's midcourse anti-missile technology failed to meet U.S. standards. He said that to be honest, at present Chinese anti-missile system only had initial combat capability.

At the reporter's question of what the military meant by saying, "The test attained its pre-set goal," the expert said, "It is very simple. Accurate interception and accurate destroy."

China's previous missile interception test was carried out on January 11, 2010 in the same month of the second test. Why January was selected as the month for both tests?

The expert said that in January the environment in space facilitates the test as the visibility is good. In professional terms, the month is a "window", but is not the only "window" in a year. The environment in September, late autumn in northern China is also good.

As for the close interval of one day between Chinese and American tests, the expert said that it was merely a coincidence.

According to the expert, the successful test showed that China was able to intercept the ICBMs that attacked China. There are three sections in the flight of an ICBM: the beginning, middle and terminal sections. Interception of an ICBM at the beginning section has the best result, but it is too difficult as the missile is accelerating. Midcourse interception as done in the test in question is effective and is what only the U.S. and China can do.

Terminal section interception is easier as the ICBM has re-entered the atmosphere and near the earth. In addition, there is more time for preparations. However, it is the last resort as the radioactive materials in the debris of the missile will fall into our country.

Missile interception is a systems engineering project in which every link counts. Through the current test, the overall capability of China's anti-missile system as well as PLA's active defense ability has been enhanced. China's interception is not directed at any country, but when China is attacked in the future, especially with ballistic missiles, China is able to destroy the attack missiles.

However, the expert believed that China only had initial combat capability as it was unable to destroy more than one missile simultaneously. In addition, China is not entirely sure whether it can hit the target without preparations and notice beforehand. Besides, the interception is an extremely great ordeal as to whether China is able to destroy an enemy missile in space before it hits the ground. All these require China to make further improvement.

Compared with the United States, China lags behind mainly in the following three areas: First, China does not have the ability to fully detect and identify the location of the target. The U.S. has 24 global positioning satellites to detect and observe all the targets in the world. China has weaker ability and smaller scope in such surveillance.

Second, less quick response ability. In the United States, there is a system of closed links between the president, Joint Chiefs of Staff, Defense Department at the Pentagon and the combat troops on duty, but such a system has not taken shape yet in China.

Third, China's test is not carried out under the adverse circumstances of real war without notice in advance. China intercepted the missile after being notified in advance. That is not the case with the United States.

The expert said: people outside China believe that at present China's midcourse anti-missile test meets the same standards as U.S. test, but in fact, it is not the case. China fails to meet U.S. standards.

However, why did the PLA announce China's anti-missile text so promptly this time?

The expert said: That is mainly due to the international and surrounding situations, especially U.S. pivot to Asia, which deteriorates the security situation around China. In order to effectively safeguard China's national security and protect China's core interests, it is necessary to show China's will, determination and capabilities. The anti-missile test itself constitutes deterrence. Moreover, it also makes China's weapon development more transparent.

ICBM interception is much more difficult than the interception of a satellite. A satellite has fixed orbit so that there is much time for calculation and preparations. There is little time in ICBM interception from detection of the launching, through identifying, tracking, collecting data, calculating its speed, trajectory and possible target, reporting top leader and the decision-making of the top leader. It must be done very carefully because miscalculation my lead to uncalled-for war between two countries.

At present, it takes the U.S. 32 seconds from the discovery of an ICBM to being on high alert; while it takes Russia 1 to 3 minutes. For China, it takes a little longer than Russia.

China's Third ICBM Interception Test on July 23, 2014

Like China's second successful ICBM interception test, the third test was made public by huanqiu.com, an affiliate of Chinese government's mouthpiece People's Daily. Huanqiu.com reporter says that he learnt from the Chinese Ministry of Defense that on July 23, 2014, China conducted a successful test of ground-based midcourse interception of ICBM.

As mentioned above, it was China's third successful interception test. Like the second, China declared that the test attained its pre-set goal.

In the preceding section, a strategic missile expert explained the meaning of "attained its pre-set goal", saying, "It is very simple. Accurate interception and accurate destroy."

Huanqiu.com says according to military analysts, the successful test is of great strategic significance. It was the beginning of a new stage of China's anti-ballistic missile technology in the areas of information processing, reconnaissance, early warning, interception weapon, weapon transmission, accuracy of guidance and speed of response.

It clarifies that China's midcourse ballistic missile interception is by no means similar to U.S. PAC-3 and Russia's S-400 missile systems. Midcourse interception of ballistic missile is much more difficult than terminal interception conducted by systems like PAC-3. It requires the tackling of very poor conditions of operation outside the atmosphere. There must be top science and technology for missile and weapons of space warfare including, among other things, KKV interceptor, accurate detection, tracking and terminal guidance technology, overall technology for space warfare platform and wartime measuring and control technology of the platform.

So far U.S. GMD system is the only ground-based system comparable in technology to China's midcourse anti-missile test.

U.S. Failed to Detect China's Strategic Nuclear Submarines

Strategic submarines are a country's top secret; therefore, though China has been developing its strategic nuclear submarine since 1966, there had never been direct official news about it. In May 2013, U.S. Chief of Naval Operations Adm. Jonathan Greenert told Congress that despite deploying a current force of 55 submarines, both diesel and nuclear powered, Chinese navy "is not there yet." "we own the undersea domain," he said.

That reminds me of General McArthur's arrogance in the early year of the Korean War. He believed Chinese troops with inferior arms dared not to fight his well-equipped troops that dominated the air and sea, but forgot such inferior troops would certainly try their best to avoid being detected in order to avoid U.S. air raid. It was unbelievable to him that Chinese troops were able to hide in North Korean mountains in spite of the cold weather, but he forgot that it was natural they had to as U.S. bombs are much worse than cold weather.

His arrogance caused him to lose common sense: Failure to detect something did not mean that the thing did not exist.

Greenert forgot that strategic submarine was made to avoid detection by others; therefore, it was only natural that the U.S. could not detect the strategic submarines China had made and deployed.

China continued to make and deploy strategic submarines to have a submarine fleet of 55, but none of them were able to go to the ocean to carry out vital second strike. Did that make any sense?

If Greenert had not been arrogant, he would have said, "We have never detected any activities of China's submarines. It seems they have left the entire oceans to us."

Source Close to China Began to Correct U.S. Underestimate

Two month later, the underestimate remained uncorrected. On July 23, 2013, Washington Times carried the report by Bill Gertz on The Washington Free Beacon titled "Red tide: China deploys new class of strategic missile submarines next year", stating that Type 094, China's second-generation strategic submarine, would not be deployed until 2014 and that it would "be the first time China conducts submarine operations involving nuclear-tipped missiles far from Chinese shores despite having a small missile submarine force since the late 1980s."

To correct the underestimate, the next day, Hong Kong's pro-Beijing *The Mirror* monthly revealed to Taiwan's Central News Agency in haste an article it was going to publish in its next issue, giving some shocking information about China's strategic nuclear submarines.

The Mirror's report by Liang Tianren reveals that the Jin-class (i.e. Type 094) submarine is China's second-generation strategic nuclear submarine. Along with it, there is China's second-generation Shang-class attack nuclear submarine.

After China's success in producing two Type 094 Jin-class and two Type 093 Shang-class submarines, according to the report, "China began development of its third-generation nuclear submarines, i.e. Type 095 Sui-class attack nuclear submarines and Type 096 Zhou-class strategic nuclear submarines and began producing them in 2007. The existence of the third-generation submarines is an undeniable fact.

It is said that the Zhou-class ballistic missile strategic nuclear submarine has strong ability to break out of island chains alone. The number of ICBMs it carries has risen to 16 from 12 for Xia-class and Jin-class submarines. Each submarine can carry 160 separately guided nuclear warheads.

Then China began development of its fourth-generation nuclear submarines in accordance with its policy of 'completing the earlier generation, building the present generation and researching in advance into the next generation'

According to informed source, the fourth-generation nuclear submarines are revolutionary submarines with unprecedented propulsion by magnetic liquid. It, therefore, has no screw propeller or tail or horizontal rudders and may possibly be free of a sail hull. As a result, it generates no mechanical or cruise noise at all and will be a real black hole in ocean.

In theory a fourth-generation submarines under development will have a speed of 100 knots, quicker than all high-speed torpedoes let alone being hit by them.

U.S. Failed to Find Information about China's Strategic Submarine in China's Public Media

Since Obama regards his pivot to Asia as a wonderful invention to contain China, the U.S. certainly has to monitor major Chinese media to know China's real strength as its support for Japan may well lead to a war with China.

If the U.S. had monitored China's *Oriental Outlook Weekly,* a major journal under Chinese government's *Xinhua News Agency* and made some analysis, it would not have been so ignorant of China's true strategic submarine capabilities

Crew of New-type Nuclear Submarine Awarded by Xi Jinping

Oriental Outlook reporter Shan Xue said in early September 2013 that he had looked into files and found that Sub-unit 90 of Unit 92730 rewarded a first-class merit by Xi Jinping in late August, 2013 is the crew of China's new type of submarine in South Sea Fleet.

Chinese military website posted a set of photos in 2010 on the return of Unit 92730 of the South Sea Fleet after satisfactory fulfillment of a major task.

Digging deeper, he found a report in PLA Daily in September 2008 on the said Sub-unit 90 of Unit 92730 being denominated a model unit for other units to learn from.

Hu Zhongming, commander of the sub-unit ranked the first among the models to be learnt from in entire PLA. From another report, it is found that Hu was the captain of a new type of submarine with the rank of senior colonel.

That new type of submarine was commissioned in February 2004. The report says, "compared with old equipment, the new-type submarine contained improvements by leaps of bounds in the materials it is made of, configuration of equipment, performance of its weapons and equipment, and extent of automation." Another characteristic is its heavy tonnage, which makes it more difficult to control.

If a U.S. submarine expert had read that piece of news, he would certainly have been aware that China began to deploy a new type of submarine on or before February 2004. Judging by the difficulty in controlling it due to its heavy tonnage, it must be a strategic nuclear submarine.

What is especially worth notice is its captain's high rank as a senior colonel because most submarine captains in Chinese navy are but lieutenant colonels.

Few captains of Chinese warship are of such a high rank. Even the captain of China's first aircraft carrier the *Liaoning* is but a senior colonel.

Official Disclosure of China's Strategic Nuclear Submarines

However, when there might be a war between Japan and China probably involving the U.S., Chinese top leaders were worried about possible U.S. nuclear attack due to U.S. underestimate of China's second-strike capability.

First, on August 19, 2013, one of China's top official media xinhuanet.com revealed the successful development of China's fourth-generation nuclear submarine in its report titled "Official news: China has successfully developed such equipment as fourth-generation nuclear submarine and aircraft carrier". It says: Tan Zuojun, Vice Governor of Liaoning Province mentioned Liaoning Province's achievements in the past decade at a conference on cooperation among the four Northeast provinces and region being held at that time. Talking about Liaoning's achievements in developing major hi-tech equipment, he said that Liaoning had successfully produced such equipment as 1,000 MW nuclear and thermal generating sets, one million-ton ethylene production equipment, high-speed locomotive, advanced ships, marine drilling platform, sophisticated numerical control center and heavy numerically controlled machine tools. He especially mentioned Liaoning's tremendous achievements in developing defense technology and industry, citing as examples the successful development of fourth-generation nuclear submarines, aircraft carrier and new-generation of fighter jets.

He has thus confirmed Liang Tianren's report on Hong Kong's pro-Beijing *The Mirror* monthly about China's development of its fourth-generation nuclear submarine.

As the U.S. may fail to notice xinhuanet's report, in order to conclusively convince the U.S. about China's second-strike capabilities, from October 27 to 29, 2013, China's top official media CCTV displayed in its primetime news footages about China's strategic submarines. The submarines displayed in the footages are China's first-generation strategic nuclear submarines soon to be decommissioned, but the footages showed China's confidence in its capabilities.

Since then there have been some appearance and news of China's strategic nuclear submarines.

During the Chinese Lunar New Year in February 2014, three type 094 strategic nuclear submarines were seen at Yalong Gulf South Sea Naval Base in Sanya, Hainan Province. The Photos taken and posted on the Internet by a web user has been much reposted.

China Reveals Its Development of Magnetic Liquid Propulsion

Dr. John, NSL (Navy Submarine League) Asian navy development advisor, said: China has now begun development of a new naval active/passive torpedo confrontation system (NAPTC) similar to that used by German Type 214 submarine. According to recent information, the technology may possibly come from German navy. It must be a kind of secret transaction between German and Chinese militaries unknown to German Chancellor Angela Merkel. As it runs against her China policy, she might have been infuriated if she had known that.

An official of China's navy research institute has recently said, "China is drawing up a plan for its fourth-generation conventional submarine, which will use magnetic fluid water jet propulsion technology. It is a technology that Europe has just begun initial application. That indicates that China has begun to follow more closely the trend of international submarine design. However, that being the case, U.S. navy will face new ordeal.

According to Dr. John, it is the first time that PLA new submarine will use European advanced submarine technology. It proves that PLA is improving its capability in designing and manufacturing submarine at flying speed.

He said: "Not long ago, Chinese navy bought from Russia a number of submarine launched anti-ship cruise missile to be used by their Yuan-class and improved Song-class submarines. But the missiles are used for tests mainly in their Yuan-class submarines.

Using magnetic fluid water jet propulsion will greatly enhance their submarines' ability to survive underwater and resist anti-submarine detection. At present, China is able to reduce to the minimum the probability of being detected for their Song-class submarines. An entirely new Yuan-class submarine, China's third-generation conventional submarine, can sink deeper and hardly be detected.

He said he was shocked when he heard Chinese expert's briefing on magnetic fluid water jet propulsion technology at NSL's International Submarine Technology Research and Development Symposium. Due to the weapon export embargo implemented by the EU against China, he personally believes that China has the technology on its own due entirely to its own invention and creation.

China Building 2 Most Advanced 098 Strategic Nuclear Submarines

After disclosure of China's Types 094, 095 and 096 nuclear submarines, people outside China have great concerns whether China really has any Type 098 strategic nuclear submarines. In May 2014, a U.S. media said in its report, U.S. satellite had detected China's construction of 2 new-type fourth-generation strategic nuclear submarines, each of which is equipped with 12 sets of launching devices for ballistic missiles.

It is said that a Chinese Type 098 strategic nuclear submarine has a displacement of 28,300 tons, can submerge to the depth of 5,200 meters and is wide in size. As it uses superconductivity electromagnetic propulsion system without any propellers, it is a silent submarine with noise less than 20 db and it can suddenly accelerate to a very high speed, quicker than any advanced high-speed torpedo. There are quite a few vector propulsion devices at various locations of its body to give it great maneuverability at various angles. Due to its low noise, it is able to hide in the complicated background of ocean noise without being detected. It is, therefore, a real intelligent undetectable stealth submarine.

It has 50 torpedo launching tubes for conventional defense while its major weapons are the 24 new JL-5 SLBMs with 80 separately guided warheads each, in which 20-35 are decoys. However effective the enemy's missile defense systems are, it can be absolutely sure that 84 enemy cities will be destroyed at the first round of attack.

There is another speculation that by the year 2025, China will have 6 strategic nuclear submarines. Each submarine carries 12 SLBMs with 3 nuclear warheads each. Two of them will keep routine patrol while a third of them will be deployed when there is an emergency. The 108 nuclear warheads they carry will constitute effective deterrent against the U.S.

China's Magnetic Liquid Submarine Will Dominate Its Surrounding Seas

In December 2013, Chinese navy launched a new conventional submarine that has drawn widespread attention. Some foreign media guessed that the submarine used the most advanced superconductive magnetic liquid design similar to the most advanced German submarine in service.

U.S. Global Security and Japanese World Warships magazines believe the photo of the new submarine shows that the design scheme of Chinese Navy's new superconductive magnetic liquid submarine has passed assessment and research, development and production for it will soon begin.

The above are all speculations.

According to the information that has been made public in China, the Electronic Engineering Research Institute of China Academy of Sciences has been stepping up its research of large-power superconductive magnetic driving device for high-speed warships. The above-mentioned U.S. magazine believes that it reflects that the PLA has a plan to build such a submarine so that the institute is developing the superconductive magnetic liquid driving device.

A new submarine propelled by such a device is characterized by zero noise, deeper submerging capability, high speed and greater maneuverability.

According to German *Naval Science and Technology magazine's* speculation, the U.S. and Germany are now leaders in superconductive magnetic liquid technology and will turn out their experimental prototypes by 2017. As it takes 3 to 4 years to succeed in research and development of such a submarine, China is quickly catching up.

9. Aircraft Carrier & Aircraft Carrier Killers

China Is Building on Its Own an Aircraft Carrier to Be in Service before 2017

An article on the website of UK's Jane's Defence weekly pointed out that China's first homegrown aircraft carrier is being built in Dalian and is expected to be commissioned before 2017

At the same time, Canada's Kanwa Defence Review also disclosed that it had finally learnt that construction of China's first homegrown aircraft carrier was being carried out in Dalian.

According to Kanwa, there has finally been a decision to grant the contract on the construction to Dalian Shipyard, putting an end to its 4-year competition with Shanghai's Jiangnan Shipyard for the contract.

However, what Dalian has obtained is the master contract. As the construction is a huge project, there are over 2,000 subcontractors, each of which is in charge of some parts of it. Dalian Shipyard is in charge of the major part including the hull and the final assembly of the vessel.

Jiangnan Shipyard has certainly got some of the subcontracts but failed to get the master contract because Dalian has the experience of the transformation of a hull form Ukraine into a workable aircraft carrier, the *Liaoning*. It is said that when Ukraine sold the hull to China, due to U.S. interference, it had dismantled the internal partitions meant for the carrier so that Dalian had to design and build the partitions that Ukrainian experts found satisfactory later.

Source said that the new carrier would not use catapult, which may be one of the difficulties China has to overcome in producing its aircraft carriers. However, there was a previous report that wide thick steel plates and the welding of them are also tricky problems.

On June 4, 2014, China Science Daily says in its report "Production of steel billets for wide thick plates has to be resolved" that China has the steel rolling equipment for the production of wide thick steel plates but cannot produce the huge steel billets to feed the rolling mill.

The demand for various huge steel billets is large as an aircraft carrier needs 40,000 tons of various wide thick steel plates.

Has this problem been resolved? Only Ukraine can help China resolve the problems, but it is in chaos now. It is said that China has attracted quite a few experts from Ukraine as it is rich now and can afford the high salaries.

Anyway, it remains a mystery to me how the plates become available for China.

An aircraft carrier without catapult cannot carry a large fixed-wing AEW&C aircraft. It needs the protection of the AEW&C from airfields along Chinese coast. As a result, it is only useful in the sea areas near China and cannot protect China's trade lifelines in the Oceans. For air battles with Japan, aircrafts from coastal airport have enough range; therefore, China needs such an aircraft carrier only for operations in the South China Sea. However, if China builds an air base on a reef in the middle of the area within the nine-dash line, it can control the sky above the South China Sea. There is no need for an additional aircraft carrier.

Due to the above reasons, I don't quite believe the reports by Jane's and Kanwa.

There has been quite much speculation about China's plan to build nuclear aircraft carriers, but I believe that China shall never build any nuclear aircraft carrier.

China Shall Never Build Any Nuclear Aircraft Carrier

As mentioned above, in missile era, as missiles are fierce attack weapons and very difficult to defend, a large aircraft carrier constitutes an easy huge target moving at low speed for missile attack. To defend, the carrier has to hit missiles small in size and moving at high speed. The odds are obviously on missiles.

Moreover, a state-of-art aircraft carrier is very expensive to build and maintain, but is vulnerable to the saturate attack of the state-of-art ballistic and cruise missiles that cost much less. The era of aircraft carrier is over and missiles are dominating weapons now. However that does not mean that U.S. aircraft carrier is useless at all.

It is useless before China's state-of-art anti-ship ballistic and cruise missiles within their range of less than 2,000 km. However, so far only China has developed such medium-range anti-ship ballistic missiles. U.S. aircraft carriers remain formidable for other countries.

I am against Chinese development of nuclear aircraft carriers not because of the carrier's inability to deal with saturate ballistic and cruise missile attack. It is an issue of principle.

The U.S. wants its state-of-art aircraft carrier very expensive to build and maintain not because it is ignorant of its inability to deal with saturate missile attack but because it wants to maintain its world leadership. Such leadership,

110

unfortunately, is regarded by some people as world hegemony. Since China has repeatedly made clear for decades that it does not pursue hegemony, why shall it build nuclear aircraft carriers? Because it has too much money and nowhere to spend? No, it needs lots of money to overcome pollution, build and improve infrastructure and improve people's living standards.

True, there are quite a few leftists who want China to replace the United States as number one in the world. Senior Colonel Liu Mingfu's book "Chinese Dream" on China replacing the U.S. as number 1 militarily soon sold out. It proves that there are quite a few Chinese people who advocate Mao Zedong's sinocentric idea. However, as described in my book *Tiananmen's Tremendous Achievements Expanded 2nd Edition*, China is now run by the new generation of talented scholars with moral integrity. They certainly are wise enough not to take over from the U.S. the heavy burden of world leadership even if the U.S. wants to.

Now, the U.S. does not want China to replace it. American people applauded when their president claimed that the U.S. would remain world leader for 100 years to come. The fight for leadership may well lead to a war between the U.S. and China that may well trigger World War III.

Some Chinese people argue that China needs nuclear aircraft carriers to defend its trade lifelines. That is a false argument. It takes China perhaps three decades to catch up with the U.S. in developing and operating nuclear aircraft carrier battle groups equal to those of the U.S. who has experience in building and operating aircraft carriers for more than seven decades.

What China can rely on during the three decades?

Nothing if China confronts the U.S.

China shall maintain good relations with the U.S. and avoid U.S. suspicion that China has any intention to replace the U.S. as world leader. By the construction of a nuclear aircraft carrier, China will precisely get the opposite. It will turn the U.S. into its enemy instead of the world police that protects China's trade lifelines.

China shall switch its financial resources to its space program to explore first the moon and then Mars. Exploration of Mars requires high speed 10 or 20 times the speed of 11,000 meters per second achieved now. It will reduce the travel to Mars from 7 months to 10-20 days. Certainly, there are lots of very difficult problems to tackle but it is what the human race has to achieve. The most difficult problem is to achieve high speed and control high-speed spacecraft to make it slow down and land on Mars and the earth.

The development of such technology will enable China to have hypersonic weapons to protect its trade lifeline perhaps within 2 decades, much earlier than the development of a nuclear aircraft carrier fleet. As mentioned in Chapter 3, China may even develop flying vehicles quicker than nuclear ICBMs so as to intercept and destroy ICBMs in midcourse and forever remove the threat of nuclear holocaust to the human race.

China Building 12 Super Warships to Be Used as Aircraft Carrier when It Has VTOL Fighter Jets

According to Jane's Defense Weekly, China will build 12 super warships with displacement exceeding 10,000 tons each, including six 071 landing platform docks (LPDs) and six 081 LPDs for helicopters. What a big spending.

China has three amphibious assault ships of the Type 071 class, the *Kunlunshan, Jinggangshan* and *Changbaishan*. The three ships are designed to transport and land marines on enemy shores. Three more of them will be built. In addition, China will build six amphibious ships with full-length flight decks.

A Type 071 LPD with 20,000-ton displacement is about 213 meters long. It can transport up to 800 marines and 18 armored vehicles.

It can carry 6 Z-9 helicopters, 2 on deck and 4 in its hanger. On its long well deck, it carries and can launch four China-made hovercrafts for marines.

The ships have a flight deck capable of simultaneously operating two W-9 troop-carrying helicopters, and can store another four in a large hangar. The ships also have a very long well deck that can store and launch hovercraft amphibious vehicles

China's 071 amphibious transports are based with China's South Seas Fleet, where they can be used to intimidate—or invade—Taiwan. However, like Western navies, China has been quick to embrace their use in other roles. In addition to assaulting coastlines and islands, the LPDs can also serve in the command and control, disaster relief and humanitarian assistance roles. One transport, Jinggangshan, has taken part in the Malaysia Airlines Flight 370 recovery effort in Indian Ocean.

As for 081 LPD for helicopters, Jane's only reveals: Construction of the first 081 LPD has already begun. According to an Asian source, the ship is similar in size to 071 LPD but equipped with better and more powerful air defense systems. Before 2020, China will initially establish its amphibious force along with as many as 4 new-type aircraft carrier battle groups.

However strong China is, China shall not replace the U.S. as world police and shall not invade any country to interfere with its interval affairs; therefore, China need no nuclear aircraft carriers. A fleet of aerospace bomber will be more powerful in defending its trade lifelines. It will be easier and take less time to develop and build than nuclear aircraft carriers.

China's navy buildup including the two conventional aircraft carriers it will build is for its operations to deal with Taiwan and the maritime territorial disputes in East and South China Seas. For that, in addition to Type 071 LPD and Type 081 helicopter carriers, it has imported four Zubr hovercrafts from Ukraine and will build some more using Ukrainian technology.

There is no need form nuclear aircraft carriers though there has been news that some Chinese research institutes are developing small nuclear reactors. Such reactors can be used by Chinese spaceships and even aerospace bombers.

China's Aircraft-Carrier Killers

In late October, 2013, the website of Japanese *The Diplomat* magazine disclosed top confidential information about China's DF-21D aircraft carrier killer. It said that the missile could destroy low-speed target with accuracy of tens of meters.

The news has shocked U.S. military as such accuracy is enough to hit a U.S. aircraft carrier which sails at the low speed of about 30 knots and is more than 300 meters long and 75 meters wide.

The magazine's report is quite authoritative as it was written by Andrew S. Erickson, a top expert in that field.

On January 12, 2014, a photo of suspected test of the launch of DF-21D anti-ship missile was posted on the Internet. Judging by the photo, the warhead of the missile is cone-shaped and there are nozzles on the missile for adjusting its direction. Perhaps the nozzles are used for terminal guidance and substantial maneuver to break through enemy defense. That is the first exposure of the test launch of the latest version of DF-21D after the first exposures of China's DF-21A and DF-21B missiles.

According to an article by Wang Genbin, deputy commander-in-chief of Department 4 of China Aerospace Science & Industry Corp. (CASIC), on a journal publicly available, in the two decades since 1988, China spent 3 billion yuan ($494 million) in successfully developing DF-21A, 21B, 21C and 21D missiles and completed the transition from development of only nuclear missiles to that of both nuclear and conventional missiles and from fixed target to low-speed target. In addition, the accuracy has been improved from several hundreds to several tens of meters.

In an interview with Asahi Shimbun, former U.S. Pacific Commander Adm. Robert F. Willard said that China's DF-21D has acquired initial operating capability

According to a U.S. research institute, in 2011 and 2012, China conducted quite a few launches of DF-21D in the South China Sea and successfully hit and sank a simulated model of aircraft carrier made by transforming China's Yuanwang 4 survey ship.

In the period from 2012 to 2013, China for the first time deployed for combat its DF-21D anti-ship missiles. One combat brigade of such missiles is located near Zhaoqing, Guangdong Province while a second brigade will be deployed in Anhui or Henan province. At present, U.S. military believes that China's anti-ship ballistic missile has already had real combat capability.

YJ-12, Another Carrier Killer When Coordinated with DF-21D Missile

In October 2013, U.S. Strategy Page website disclosed China's another aircraft carrier killer YJ-12, which is second in power only to DF-21D ballistic missile.

It is an advanced high-speed anti-ship missile with a range of 400 km able to break through all Western short-distance interception systems in service. Its speed reaches Mach 4 at the terminal stage.

YJ-12 is 7 meters long, a little smaller than Russian 3M55 and 3M80 missiles; therefore its launch weight may be a little smaller, but its launch weight may still be relatively big as when it has discarded its booster, its mass at the terminal stage are relatively big at somewhere between 2 and 2.5 metric tons.

According to military experts, YJ-12 is extremely accurate as it is guided by both the Beidou Satellite Navigation System and a terminal broadband active radar system. The missile first climbs to a certain height. When its data link has received confirmation of the parameters of its first target from the early warning radar, its guidance system sends the parameters to the flying control system and the missile begins to dive down to cruise at low altitude.

At that stage, though it is slow at Mach 1.5, it is difficult to be detected as it cruises silently 12 to 15 meters above water under satellite navigational and inertial guidance. When the missile is within 50 km of its target, the active guidance radar is switched on to confirm with the satellite guidance the newest parameters of the target and then compare such parameters with those in the data link. After confirmation of the parameters, the missile enters the terminal attack stage at the speed of Mach 4 difficult for short-distance interception.

According to analysis of its purpose by industrial sources, YJ-12 uses liquid ramjet most probably because in theory liquid fuel takes lots of air, generates high energy and adjustable thrust and is suitable for large variation of height and speed and long range.

If YJ-12 is carried by a Flying Leopard fighter-bomber with a range of 1650 km, with its own 400 km range, its range of attack will be 2050 km.

If it is carried by China's Su-30MKK, J-16 and H-6G/K with longer range than Flying Leopard, its range will be even longer.

British *Jane's Defence* Weekly believes YJ-12 anti-ship missile can be launched not only from an aircraft but also from a warship, submarine and land launch vehicle. If coordinated by a long well-known DF-21D missile, it will enable China to have coordinated attacking capability from both high and low altitude, making it difficult for enemy warships, especially aircraft carrier to defend.

DF-26 2nd-Generation Aircraft Carrier Killer

According an issue of U.S. Aviation Week & Space Technology magazine in January 2014, U.S. Navy regard the Mach 10 hypersonic glide vehicle (HGV) China tested on January 9, 2014 as an omen of the appearance of an anti-ship ballistic missile (ASBM) much better than DF-21D to be developed by China using the HGV.

At the same time, U.S. *Aviation Technology and Space Weekly* carries an article in its January 17 issue titled "U.S. navy regards China's hypersonic glide vehicle as the part of Chinese weapons with extensive threat", pointing out according to U.S. navy, the Mach 10 HGV China tested on January 9 reflects China's foresight on future war. Once China is able to apply that technology, it will have a weapon that can challenge all existing missile-defense systems and widen the range of its ballistic missiles. It takes a few years for such weapons to be usable depending on resolution of the difficult issues of controlling its guidance and making it hit accurately.

The report says that the HGV test makes a step forward of China's research into anti-ship ballistic missile (ASBM) and probably signals the appearance of China's second generation of ASBM. According to U.S. China military expert Richard Fisher, the DF-26 missile mentioned in rumor perhaps has a HGV warhead and will have a longer range of 3,000 km than DF-21D's 2,000 km. If China's DF-31 ICBM is installed with such a warhead, its range will be lengthened from 8,000 km to 12,000 km.

The report believes that the U.S. has to develop directed energy weapon to counter such missiles as the existing missile-defense systems cannot intercept missiles with a speed exceeding Mach 5.

Fleet of Type 022 Fast Missile Boats, China's Another Aircraft Carrier Killer

Since 2004, China has deployed over 80 Type 022 fast missile attack boats (also called Houbei class fast boats).

The boat has such outstanding capabilities that John Patch, a retired U.S. Navy officer said in his article for the United States Naval Institute in 2012, "This craft is a purebred ship killer, perhaps even a carrier killer".

Britain's Jane's Defence weekly believes China's Type 022 stealth fast missile boat is perhaps the first stealth fast missile boat in the world that adopts wave-piercing technology. Its three great advantages of high speed, quietness and undetectability make it so powerful that under certain circumstances, it may scare away a U.S. aircraft carrier.

Its shape with lots of sharp angles greatly reduced the reflection of radar wave.

Its water jet propulsion avoids the noise caused by the air bubbles in propeller propulsion to make it very quiet. A photo of its water jet can be viewed at http://tiananmenstremendousachievements.files.wordpress.com/2013/10/water-jets -of-type-022-fastboat.jpg

On October 28, 2013, a reporter of Chinese Navy's website navy.81.cn was on board of a Type 022 missile fast boat to watch the drill of a flotilla of Type 022 fast boats. According to him, the boats' stealth function is so superb that tens of radars cannot detect them while their coating is so wonderful that one cannot see them until they have come quite close. Photos of the drill can be views at http://tiananmenstremendousachievements.files.wordpress.com/2014/04/drilling-o f-chinas-022-fast-boat-fleet-regarded-as-aircraft-carrier-killer.jpg, http://tiananmenstremendousachievements.files.wordpress.com/2014/04/drilling-o f-chinas-022-fast-boat-fleet-regarded-as-aircraft-carrier-killer-2.jpg, and http://tiananmenstremendousachievements.files.wordpress.com/2014/04/drilling-o f-chinas-022-fast-boat-fleet-regarded-as-aircraft-carrier-killer-3.jpg.

Jane's weekly said in its report that the boats' eight YJ83 new-type medium-range (150 km) anti-ship missiles can attack multi-targets beyond visual range. A flotilla of more than two dozens such boats can conduct saturate attack by simultaneous fan-shaped launch of all the boats' missiles.

That is quite formidable as the boats can approach enemy fleet to a very short distance without being detected due to its radar and optical invisibility and quietness. They may leave the battlefield after firing their missiles as due to the boat's advanced data link, they can leave the missiles to the care of Chinese AEW&C aircraft or other warships.

Each boat has air defense missiles and a 30mm six-barrel high-speed gun for short-range missile defense.

The boat's multi-purpose radar can search and lock on targets both in the air and at sea and guide anti- ship and aircraft missiles. It has also optical-electronic and infrared tracking devices and laser distance measuring device. In addition, it can provide navigation guidance and short-range early warning for other boats that are maintaining radar silence.

The boat uses the wave-piercing catamaran structure that China leant through its joint venture with Australian high-speed ferry designer AMD Marine Consulting established in 1993.

The Chinese joint venture partner Guangzhou Marine Engineering Corporation, a subsidiary of China State Shipbuilding Corporation, gained access to the state-of-the-art technology in wave piercing, aluminum-hull designs.

Attracted to the performance of these fast, stable and relatively cheap vessels, the Chinese military adopted the technology as it began to develop an advanced fast boat to replace its aging missile boats of obsolete Soviet design.

The fast boat has a light aluminum double hull with a displacement of only 225 tons that enables it to have an estimated top speed of more than 36 knots.

The boat is a clear example of the value of foreign dual-use technology in China's rapid military buildup. It obviously is adapted from the AMD's patrol boat design.

However, AMD's technical director, Allan Soars, said the Australian company was not involved in the design of the missile craft.

Then who designed the boat?

A Pretty Girl Yang Yi Designed China's Type 022 Missile Fast Boat

According to Chinese official media mil.huanqiu.com's report in November 2012, the boat was designed by a pretty girl named Yang Yi.

When she was interviewed by a reporter in 2012, she may be 40, but still looks young in a photo provided by Chinese navy at . However, nearly 2 decades ago in 1994, she was appointed chief designer of China's well-known Houbei Class fast attack missile boat. At that time, she was the youngest chief designer ever in China.

When she was appointed, fast boats seemed outdated. Perhaps, that was the reason to entrust a young girl the design of a new type of them, though she had already made prominent achievements in fundamental research.

Through 8 years of hard work, she finally succeeded in developing that very advanced stealth fast boat able to operate at high sea. She won a PLA scientific and technology progress award for that in 2002.

The boat's wave-piercing catamaran was learned from an Australian company, but all the other wonderful functions and performance have been designed by Chinese engineers under the young chief designer's leadership. It proves Chinese engineers' creativeness in exploiting other countries' technology to develop advanced weapons other countries do not have the least idea of.

Before the emergence of Yang's fast boats, China used its fast boats for defense of its coast. Now however, Yang's talents have turned what was for defense into a fierce weapon of attack far away from Chinese coast. The wave-piercing design enables the boat to operate at high sea.

From Yang's success, we see the mystery of China's success. As mentioned in my book *Tiananmen's Tremendous Achievements Expanded 2nd Edition*, "China's Prosperity Was due to Its Talented Intellectuals with Moral Integrity". With the fear created by Tiananmen Protests and the assistance of a new generation of talented intellectuals with moral integrity emerged during the Cultural Revolution, Jiang Zeming has successfully carried out a silent peaceful coup d'état to replace intellectuals' dominance for uneducated workers' and peasants' dominance of the Party and state. That enables Chinese reformists to give play to the talents and diligence of Chinese people, especially intellectuals.

That is a long story that has to be described in details in the book.

Under such circumstances, Chinese military leaders' appointment of a talented girl as chief designer, though exceptional, is normal.

If Xi Jinping succeeds in giving further play to the talents and diligence of China's huge population, there will be no rivals to China in the world.

10. U.S. Allies Help China's Military Modernization

Import of Dual Technology Facilitates China's Rapid Military Buildup

There is no doubt whatsoever that Yang learned from AMD Marine Consulting's wave-piercing twin hull design that enables her boats to sail into rough sea at high speed. The Australian firm provided China with the technology for the construction of ferries for civil use. However, the technology can also be used for military purpose.

Certainly, China shall also be capable of using foreign technology and combine it with what China has developed on its own. The boat is special in its electromagnetic and optical stealth coating that make it invisible and not detectable by radar. Its waterjet propulsion (which can be viewed at http://mil.huanqiu.com/photo_china/2013-10/2711799_2.html) makes it very quiet.

According to GlobalSecurity.org, the waterjet technology has probably come from Swedish company Kanwa, a subsidiary of UK's Rolls-Royce plc. It, however, also says, "Chinese military's *Liberation Army Daily* reported that 'China Water National Key Laboratory of acoustic vibration casing vibration and noise reduction technology research made fruitful achievements in scientific research.' Military experts believe that this technology is important for high-speed vessels."

Due to Chinese military's habitual lack of transparency, we really do not know how much technology in the boat has come from abroad. We only know that China has combined various imported technology with its own satisfactorily to develop a fast attack boat of world-advanced standards.

GlobalSecurity.org says, "The most noteworthy is that all 022 missile boats are equipped with one-way hn900 tactical data link, which can work together to achieve the intent. It can collaborate with other frigates, missile destroyers, shore-based reconnaissance, early warning aircraft, boats and even between the implementation of the 'proper reach,' the distant sea offensive operations." (GlobalSecurity.org's report on Type 022 fast boat can be viewed at http://www.globalsecurity.org/military/world/china/houbei.htm)

There was a drill of a Type 022 fast boat flotilla last October. According a reporter of navy.81.cn who was on board a fast boat to watch the drill of a flotilla of the fast boats, the boats' stealth functions are so superb that tens of radars cannot detect them while their camouflage painting is so wonderful that one cannot see them until they have come quite close.

In the drill, the flotilla stealthily sailed to and stayed on the route of a fleet to ambush it. The fleet conducted electromagnetic interference but the flotilla responded with its anti-interference capability. When the fleet is within attack range of the flotilla, 4 fast boats suddenly accelerated to full speed towards the fleet. The reporter saw two huge waterspouts at the tail of each fast boat.

It is said that when it attacks, its can sail at 45 knots while when it leaves after firing its missiles, it can go at 55 knots with reduced weight.

In the drill, when the commander said, "The targets have come within the best range," and give the order to fire, the doors on the stealth weapon cabins of the four boats opened simultaneously to launch several missiles. Soon the commander received reconnaissance information that the missiles hit the targets and seriously damaged the fleet.

Russia and EU, China's Important Sources of Arms and Military Technology

Russia is the major source of Chinese weapons. Most of China's advanced weapons including Su-27, Su-30, J-10, J-11, J-11B, J-15 fighter jets, S-300, S-400, HQ-9 air-defense systems, etc. are either imported from Russia or designed on the basis of Russian weapons. China has learned a lot of technology from imported Russian weapons. Over the past few decades, it gradually became able to improve imported weapons, develop its own weapons on the basis of imported Russian ones and is finally able now to develop some advanced weapons entirely on its own.

However, other Western countries, especially EU members have given China quite much help in weapon development.

Voice of Germany says Chinese air force relies on the helicopter designed by France while Chinese submarines, destroyers and frigates have to use German and French engines.

The radio station quotes Chief editor of *Kanwa Defense Review* as saying, "European export of weapons is very important for China. Without European technology, Chinese navy can make no progress."

There has been a ban on weapon sales to China since 1989, but each EU member has its own right to interpret the ban; therefore, in the decade up to 2012, EU weapon manufacturers got licenses for export of weapons to China worth 3 billion Euros. In 2012, EU exported 170 million Euros of weapons to China, of which 80% has been French export. French parliament said in its report that

France exported 104 million Euros of weapons, most of which is export of military helicopters.

The problems are the export of goods that can be used for both military and civil purposes. They are not banned when China says they are imported for civil purposes but China actually used them for military purposes. That is the case of EU export of helicopter and submarine and warship engines.

In fact, Europe has implemented the ban loosely. That has given rise to tension between Europe and the United States.

Average annual increases of almost 12 per cent in military spending over more than two decades have allowed China to deploy an expanding force of potent warships and submarines, long-range strike aircraft, missiles and modernized nuclear warheads.

Early in this period, China relied heavily on imports of Russian weapons but this has slowed as the domestic arms industry gears up to build more locally designed hardware.

As part of this shift, dual-use technology from abroad has been crucial to advances across a broad range of China's military technologies including satellites, communications networks, helicopters, radars, marine engines, signals processing and training simulators, military analysts say.

China's state-owned commercial shipbuilders, who also deliver warships for the navy, have been at the forefront of absorbing foreign technology.

In its report on China, the Pentagon said it would continue with efforts to block the transfer of important technology to China that would contribute to China's defense industry and military firepower.

However, for the United States and its allies, it could be difficult to evaluate which technologies or materials should be restricted, according to military analysts, particularly for countries that benefit from close trading relationships with China.

Germany Enables China's Conventional Submarine to Play Nuclear Sub's Role

A media outside China pointed out in early March, 2014 that German navy has transferred to China the technology of its newest U-214 submarine without government approval. The technology has enabled Chinese submarines to surpass U.S. and Russian ones overnight.

Chinese submarines using that technology can patrol under water at the speed of 2 to 6 knots for more than 3 weeks as it uses a fuel battery system. If it uses a hose to draw air in from the surface, it can cruise at the speed of 6 knots for 12 weeks to cover 12,000 nautical miles. Such capability to stay underwater for a long distance is as good as that of a nuclear submarine. It is, therefore, superior to that of all the conventional submarines other countries have.

China has had a good mastery of such technology now and is using it to improve its existing AIP submarines. The AIP system in U-214 submarine is much better than Chinese one as it is so far the most advanced submarine technology in the world. It is better concealed as its modular structure has integrated the weapon systems and sensors with the submarine's platform. With that technology, China can use conventional submarines to do what was previously done by its expensive nuclear submarines.

The submarine's smooth body, stealth coating and very low noise make it perfectly undetectable. It cannot be detected even by magnetic and infrared detecting devices as the submarine is so strong that it can sink deeper than 400 meters, a depth difficult to reach by other countries even Japan.

This kind of conventional submarine is called "green nuclear submarine" as it is capable of cruising under water for as long as a nuclear submarine but it does not incur the risk and high cost of a nuclear submarine.

China Upgraded All Its Yuan Class Subs with German Technology

On July 14, 2014, the website of Japan's *The Diplomat* magazine publishes a report on the arms race in Asia caused by China's military buildup. It described South Korean, Japanese and Vietnamese efforts in developing their submarine fleets as it believes that submarines are a vital part of navy.

China's submarine fleet is of course what the website is most interest in. The website believes that since 2004, China has launched 12 Type 041 Yuan class conventional submarines. The submarines built earlier are certainly not advanced enough due to China's lack of advanced technology at that time. In addition due to the progress in technology, even those advanced at the time when they were built have become outdated.

China has upgraded all the 12 Yuan class submarines it has built by installation of more advanced sonar and weapon systems, reduction of noise and adoption of air-independent propulsion (AIP). Moreover, British Jane's Defense Weekly's reported on April 9 that China had launched a new Yuan class submarine that adopted the newest design of German SSK submarine. Like German SSK submarine, it has a slant sail hull, at the base of which an additional high-frequency sonar can be installed. There are other things similar to SSK.

Obviously, China has also used what it has learnt from German SSK submarine to improve its 12 existing Yuan class submarines.

In addition, China has purchased 12 Kilo class submarines from Russia and is discussing with Russia on the purchase of 4 Lada class submarines.

The website's report says that China has developed its fleet of advanced conventional submarines and attack nuclear submarines for its anti access/area denial strategy, which is referred to by China as "anti-interference" strategy.

That shows that The Diplomat does not really understand China. There have been lots of talks in China recently about resisting enemy invasion in the high sea so that the area near China will not be a battlefield that will bring damages to China.

Perhaps, *The Diplomat* knows China's intention to defeat its enemy away from its coast but bases its view on the perspective of a war between China and Japan. That is why it does not mention China's nuclear submarines in its report as those submarines are mainly used far away from China.

However, why are China and Japan unable to be friendly neighbors? True, there were two brutal wars between the two countries in the past but why shall China settle account for the crimes committed by current Japanese people's great grandfathers. Better let bygone be bygone and regard China's military development as directed at the enemy from far away.

China to Get Best French Weapon–Newest Mini Submarine

According to Japan's *Sankei Shimbun*'s exclusive report, France is trying to sell its newest SMX-26 mini submarine to both China and Japan. The newspaper says that it shows France's evil intention to help China while ignoring Japan's needs.

Japan does not need SMX-26 due to the sea around it is quite deep for its submarines and Japan has the best technology and equipment to detect submarines.

SMX-26 is in fact tailor made for China.

For China, the Yellow Sea is shallow mostly with the depth of only 40 meters. Its maximum depth does not exceed 150 meters. Almost nowhere in the East China Sea is more than 100 meters deep. Ordinary submarines have difficulties in operating in waters less than 50 meters deep and need at least the depth of 100 meters for free combat operation.

Moreover, the bottom of the East China Sea is so ragged that the location of the noise of a submarine is difficult to detect by sonar.

According SMX-26's producer DCNS Group, the submarine can operate in very shallow waters less than 15 meters deep. Its four controllable expansive azimuth thrusters provide it with very strong maneuverability and enable it to remain steady near sea bottom or surface. In addition, it is installed with landing gear with wheels to enable it to land on various kinds of sea bottom quietly. When it stays put at sea bottom, it uses the air taken from above surface by a hose.

As a result, it can remain under water in ambush for 30 days to attack enemy warships and submarines with its two heavy anti-warship torpedoes and 8 light anti-submarine torpedoes.

The report stresses that SMX-26 can be equipped with 3-dimention map drawing instrument to enable China to be clear of the morphology at sea bottom so as to decide the routes for their submarines to attack and retreat.

European countries want to sell weapons to China due to their economic recession. They now have the opportunities of sales due to Russian reduction of its weapon export to China. Now, France and other EU countries want to lift the existing ban on weapon sales to China. In doing so, they entirely fail to take into account of what Japan wants.

China's Apache WZ-10 Armed Helicopter Uses Italian Design and Technology

At the Farnborough International Airshow in the morning of July 14, 2014, lots of aircrafts gave wonderful flying performance. Among them is Turkey's T-129 armed helicopter that displayed its excellent maneuverability though it was the first time T-129 takes part in the Air Show.

T-129 armed helicopter is the export version of Italian AgustaWestland's Agusta A129 helicopter nicknamed Mongoose.

As a NATO member, Italy is located right in front of the Warsaw Pact. In the face of a flood of Soviet armors, Italy needed effective anti-tank means to contain the Soviet Union. As a result, T-129 helicopters become Italian Army's main battle helicopters. It is the first helicopter designed entirely in Europe and the first European armed helicopter that has undergone the test of real war.

A129 is quite popular on the international market. When its model was displayed at Beijing Airshow in 2001 and Zhuhai Airshow in 2004, it drew much interest among Chinese military fans.

Military observers believe there are quite a few similarities between China's

WZ-10 and Italian A129 armed helicopters. They both have a tandem cockpit with a pilot sitting higher and a gunner sitting lower before the pilot. They are similar in their locations of front infrared systems, manners of the installation of their engines and the design of their tails. However WZ-10's engine is obviously inferior to A129's. At first, WZ-10 used Canadian 1,300KW PT-6C-67C turboshaft engine for civil usage, but due to U.S. influence, Canada has refused to sell any more PT-6C engines to China. China is forced to use its homegrown Wozhou-9 engine with insufficient power.

It is known to all that China is unable to purchase armament from the West, but Sino-European cooperation related to helicopter is also known to all, for example, the Z-9 helicopter widely used in China now. It is the product of Jiangxi Changhe-Augusta Helicopter Co., Ltd, a joint venture between China's Changhe Aviation and Augusta S.p.A. It is quite interesting that Changhe Aviation played a major role in developing WZ-10 helicopter. No wonder, there are the similarities.

In a previous section of this chapter, I said, "Voice of Germany says Chinese air force relies on the helicopter designed by France."

Details of China's Apache WZ-10 armed helicopter will be given in Chapter 14

Germany, France Help China Build Its Navy at Unprecedented Speed

Canada's Kanwa Defence Review published an article in early 2014 that described China's astonishing speed in building up its navy. In 2013 alone, it launched 19 warships. It launched three 4,000-ton 054A frigates in September 2013.

China was able to build 19 such frigates thanks to German supply of advanced 16PA6 diesel engines and French supply of the technology of its Tavitac system for the frigate's HN-90 naval battle data chain system and ZKJ-4B tactic data processing system.

In fact, Chinese military have been much benefited from import of advanced European equipment.

Aircraft and warship engines take time to develop. China has so far been unable to produce most of the engines for its advanced aircrafts and warships.

Russia and Ukraine have been China's major suppliers of advanced aircraft engines, but the UK has also made its contribution. China's fighter-bombers use British jet engines.

Besides 054A frigates, most of China's advanced surface warships are powered by German and French-designed diesel engines.

China's submarine force relies even more on German engines for low noise level. It uses German company's latest diesel engine design. Its Song and Yuan class submarines both uses the state-of-the-art diesel engines designed by MTU Friedrichshafen GmbH of Friedrichshafen, Germany. China has built more than 20 submarines using German engines and imported from Russia 12 advanced Kilo-class submarines. Because of import from Europe, China has a quite strong conventional submarine fleet in the world.

However, China does not stop there. India's Hindustan Times said in its report on April 9, 2013 that according to Indian defense ministry's confidential report, China may plan to build 15 more Yuan class attack submarines using German diesel engines. As a result, the new Yuan class submarines will be equipped with air-independent propulsion to recharge their batteries without surfacing for more than 3 weeks.

In addition, China has ordered 4 Russian most advanced Lada-class submarines with noise level 8 times lower than Kilo-class ones.

Arms Export Ban Implemented Loosely

There has been an official embargo on arms sales to China since Tiananmen Massacre in 1989, but due to economic recession, EU countries want to do more business with China to improve their economy. They have imposed the embargo loosely. As a result, China has been able to get lots of weapons and weapon technology from EU. China has got the design of its advanced surface warships and superior attack helicopter from EU. Chinese destroyers are equipped with sonar, anti-submarine helicopters and anti-aircraft missiles from EU. Chinese reconnaissance aircraft is equipped with the latest British airborne early warning radar.

According to official figures, EU weapon producers have been allowed to sell $4.1 billion worth of weapons to China in the 10 years to 2011, including aircrafts, warships, imaging equipment, tanks, chemical agents and ammunition.

Michael Mann, an EU spokesman in Brussels, said the EU arms embargo issued in June 1989 "does not refer to dual-use goods." It is up to individual member states to exercise control over such goods.

France and the UK conducted the arms sale ban most generously, mostly forbidding only the sales of lethal items or complete weapons systems. France allowed almost 2 billion euros of sales, exceeding by far other countries' sales. Britain ranked second with sales of almost 600 million euros while Italy ranked the third with sales worth 161 million euros.

The actual value of arms export to China may be much bigger than the above statistics as some countries, including Britain and Germany, do not report the actual figures.

As mentioned above, the EU arms embargo issued in June 1989 does not refer to dual-use goods; therefore, the above EU arms trade figures don't include dual-use technology that in many cases can be sold without restriction. For example transfers of commercial aerospace design software are not blocked but the software can be used for fighters, bombers and unmanned aerial vehicles.

Dual-use transfers are almost certainly more valuable to China than the actual export of European weapons. But it's impossible to collect statistics of such transfers. The trade volume between China and EU is very large and there is no consistent system in EU to track these transfers.

As China is far away from Europe, the rapid growth of China's military power is not regarded as a threat to Europe. It is on the contrary regarded as an opportunity for Europe to make money.

China wants to have the embargo lifted but the pressure from the U.S. has kept the embargo in place. To avoid the restriction of the embargo, Europe sells lots of weapon components to China, especially those related to dual-use technologies.

11. Catching Up with the U.S. in Stealth Aircrafts, etc.

China's Stealth Fighter Jets

Chinese air force has been making great efforts to catch up with and surpass the U.S. in developing stealth aircrafts. There have been quite a few types mentioned by Chinese and foreign media including J-20, J-31, J-18, J-23 and J-25.

J-20

J-20 and J-31 stealth fighter jets are the two widely known types of China's fourth-generation fighter jets.

J-20 heavy stealth fighter jet is the first stealth aircraft China has been developing. Its first prototype no. 2001 had its successful maiden flight on January 11, 2011. After quite a few test flights, a second prototype no. 2002 emerged on May 10, 2012.

A second version of J-20 emerged later in January 2014. There are quite a few improvements in the second version compared with the first version. However like all the advanced aircrafts China is developing, it does not have an engine developed by China on its own.

China has allocated 100 billion yuan (U.S.$16 billion) for the development of advanced engines it needs. With such huge funds, China will be able to produce good engines in the future, especially because it has the assistance of Ukrainian experts. Still it has to wait for quite a few years as even it has successfully designed and produced a type of engine, it has to test the engine for a long time to ensure that the engine can operate safely for a long time.

As J-20 is developed to grab air supremacy from U.S. F-22, China has attached greater importance to it and there has been more information about it than any other stealth aircraft China has been developing. In late February 2014, Russian media reported that China had developed WS-15 engine with 18-ton thrust for J-20 but the engine still has to undergo years of test and improvement.

J-20's functions and performance have mostly been kept secret. I have found some reliable information from a few sources.

On January 18, 2014, huanqiu.com says in its report that according to a photo of the new version of J-20 posted at the website of a well-known Chinese military forum, compared with the old version there are the following improvements:

1. Reduced cylinder for driving the major wings;
2. Air inlet tilting down to integrate with the fuselage;

3. A yellow EOTS sensor (photoelectric target sensor) is installed at the bottom of its nose;

4. Perpendicular tail wings have been tapered;

5. Improved parachute cabin on top of the fuselage;

6. The nozzle at the tail of the engine is almost entirely shaded by its ventral fin to improve its undetectability and is zigzag-shaped;

7. Pilot cabin cover strengthened by a rib and has a prima-cord installed to enable the seat to eject through the cover in emergency;

8. The new coating that makes the aircraft less detectable;

9. Obviously smaller cover of the cabin for its major landing gear;

10. Relatively great change in its strake wings

11. The use of new diffraction flat-screen monitor with lighter weight and sharper display;

12. In the photo, we see a refuel-receiving hose on the right side of the nose suggesting China's intention to make J-20 a carrier-based aircraft as buddy refueling is the only refueling method for carrier-based aircrafts.

In an interview with China News Service on March 3, Major General Zhu Heping, Deputy Head of Air Force Command College, talked about the successful maiden flight of the new version of J-20 stealth fighter jet. What Zhu said confirmed that there had been lots of improvements including the use of better engines and enhanced combat capability and undetectability.

That proved China's great efforts in making J-20 surpass U.S. F-22 as like F-22, J-20 is designed to dominate the sky. In order to dominate the sky, it has to be better than F-22. Otherwise, there is no point to develop it in the first place.

What we so far know is that J-20 surpasses F-22 in the following areas:

First, like F-35 that emerges later than F-22, it is installed with an EO (Electro-optical) DAS (Distributed Aperture System). The DAS surrounds J-20 with a protective sphere of situation awareness so that the pilot is aware of incoming aircraft and missile threat.

Second, it is armed with PL-10, one of the best air-to-air missiles in the world in 2014. PL-10 enables J-20 to counter the threat known by EO DAS, as it is capable of rear firing at the aircraft coming to attack J-20 from its back.

There is another advantage in PL-10 that makes J-20 surpass F-22. According to U.S. *Aviators* website, unlike the missile used by F-22, PL-10 has better view when it is carried by J-20. It does not require the F-22's complicated move of locking on its target after launch by the pilot. The pilot's helmet is so advanced that when a PL-10 is launched, it will hit at the target that the pilot looks at.

As it takes time for an F-22 pilot to lock on a J-20 in a fight between the two fighters, the simplicity in taking aim enables J-20 to fire its PL-10 earlier and hit the F-22 earlier.

Moreover, due to the application of thrust vector technology, PL-10 can effectively hit a highly maneuverable target.

J-20 Equipped with Mach 5 Pili-13 Air-to-air Missile, a Serious Threat to F-22

In late September 2013, Huang Zijuan, a China's people.com.cn reporter, said in his report: According to recent report by foreign media, having conducted extensive test flights, China's stealth fighter jet J-20 has begun to carry out tests of China's newest Pili-13 missile.

Armed with Pili-13, a J-20 constitutes grave threat to U.S. F-22 fighter. Military expert Li Li said in an interview with media that if J-20 was equipped with that type of missile and assisted by an AEW&C aircraft, no aircraft locked on by them could escape the attack.

The report says that Pili-13 missile, 3.0 meters long, 170 mm in diameter and 500 mm in wingspan, exceeds in size all the existing fighting guided missiles in the world. It is similar to French MICA missile with concurrent medium-range interception function. Judging by its pneumatic shape, Pili-13 uses strake wings in its design similar in shape to Russian R-77 missile. The strake wings are located on the body of Pili-13 at the same position on the body of the missile to those on R-77. The shape of Pili-13's pneumatic rudders at its tail is similar to the reverse-trapezoid of Russian R-27 missile. It perhaps adopts thrust vector design at its tail.

The U.S. has organized air battle between fighters of the same type as F-22 and found that the air battle between stealth fighters may possibly become close-range fighting. Obviously, China has accepted that theory while Pili-13 missile is perhaps a product of that theory. If mass production of Pili-13 began soon, it could be used in coordination with an AEW&C aircraft so that the aircraft transfer through data chain to the J-11B and J-10 equipped with Pili-13 the information of the target it has detected, to enable the PLA to have the ability of super long-range attack.

According to media analysis, Pili-13 is equivalent in performance to America's most advanced Sidewinder missiles. In an air battle, no aircraft can escape when it is locked on by a Sidewinder missile. U.S. media are worried by the emergence of Pili-13. They believe that the type of missile will greatly improve the combat capability of Chinese air force and constitute a serious challenge to U.S. air force and naval air force.

Japan is in great panic. A Japanese military research monthly says that J-20 is China's most advanced homegrown fighter jet with phase array radar of equivalent performance to that used by U.S. F-22. If it is equipped with Pili-13, none of the weapons and equipment Japan purchases from the U.S. will have any superiority over them.

Regarding foreign media's allegation that if a J-20 is equipped with Pili-13, no aircraft it has locked on can escape its attack, Li Li says that she has to add one more factor, i.e. there shall be coordination between the J-20, an AEW&C aircraft and Pili-13 to make them an integrated system. J-20's own radar has a limited range, but if the AEW&C in the system has detected a target first within its range of several hundred kilometers and transfer the information to the J-20 and enable it to guide the missile to attack the target, the system will function satisfactorily.

Li Li says: many present-day missiles are of wholly intelligent design so that as soon as it has locked on a target, the target cannot avoid being hit by the air-to-air missile no matter how great maneuver it has made. Pili-13 is such a missile. That is why foreign media compared it to U.S. Sidewinder. Sidewinder is guided by infrared ray, but it is not a tailing type. It follows the trajectory of its target to hit the tail of the target. Many advanced missiles do so now. As a result, there is little possibility for the target to escape.

A photo of J-20 carrying Pili-13 missile can be viewed at http://tiananmenstremendousachievements.files.wordpress.com/2013/09/new-air-to-air-missile1.png

Intensive Flight Tests of More Prototypes of 2nd Version of J-20

In mid July, 2014, well-known Chinese military forum lt.cjdby.net posted a photo of the taxi test of J-20 no. 2012, a second prototype of the second version of J-20 stealth fighter jet. The new prototype is regarded as a further progress in developing J-20 after J-20 no. 2011 left for finalization tests at another airport in late June 2014. It is especially significant as the new prototype looks identical to its predecessor J-20 no. 2011.

The emergence of an identical second prototype months after that of the first prototype makes Chinese military fans believe that China has begun series production of J-20 prototypes.

According to information from U.S. Air Force, 11 prototypes of F-22 in all were produced for test flights in developing F-22. As there were tests of J-20's radar system on an aircraft not long ago, they believe that China will keep on producing and testing more J-20 prototypes installed with complete navigation electronic system. Such prototypes will not be much different from the J-20s that will be commissioned later.

According to U.S. Air Force, throughout the period of F-22 development and research, 2,546 test flights were conducted for a total of 4,583 hours. If J-20 has to undergo similar tests, China has to build 10 J-20 prototypes and complete all the test flights by the end of 2015.

In December 2013, Jane's Defence Review reported that China had the capacity to produce 138 military aircrafts at that time. Judging by the intensity of the arms race China is carrying out against the U.S., The capacity will grow in the future. As a result, China will be able to build enough J-20 better than U.S. F-22 to dominate the sky by 2017.

According to the web users who personally saw the test flight at the site, J-20 newest prototype no. 2012 successfully conducted its maiden flight at 1055am, July 26, 2014.

The maiden flight of the new prototype signifies another stride in developing J-20 after the first prototype of the new version of J-20 departed to another site for finalization test flights. It is perhaps the beginning of mass production of J-20 prototypes and intensive test flights of the prototypes.

J-31

J-20 is a national project funded by the Chinese government while J-31 is Shenyang Aircraft Corporation's own project funded by the corporation itself. Due to its tail hook, it is widely regarded as China's next generation of carrier-based aircraft.

As it is not a national project, there is speculation that the corporation develops the aircraft for export as due to the much lower price of Chinese aircrafts, it may grab a large market share from U.S. F-35. However, others speculate that the corporation wants the PLA to accept the aircraft when it has successively developed J-31 so that it may not only recover development cost but even make some profit.

However, if J-31 is developed for export, it will make China a rival to its ally

Russia in weapon export market. As a result, Russia will not be willing to provide the powerful engines that J-31 needs but China cannot make.

There was news in late January 2014 that J-31 will get Ukrainian engine for light stealth fighter with super undetectability.

According to speculation by such media as Russian "Armed Forces Journal" and American "Wired" monthly in early 2014, Ukraine will probably provide advanced AI222K-95F engine to enable PLA to have a light stealth fighter.

Due to the engine's satisfactory ability to hide its infrared emission, it will enable China to have a light fighter with super undetectability. In addition, with that engine, the fighter jet can fly at high supersonic speed.

The widely known J-31 is perhaps merely an experimental model. With such an engine, China will really have a light stealth fighter jet comparable to U.S. F-35 to supplement its heavy J-20 stealth fighter jet.

J-20, however, still lacks powerful engines to provide it with satisfactory maneuverability. That is why it has such big front wings to improve its maneuverability. It should also use imported engine, but Ukraine does not have any. Russia's AL-41F designed for its T-50 stealth fighter may be the one J-20 needs, but China is not sure Russia will be willing to provide China with such engines.

J-18 VTOL Stealth Fighter Jet

Japanese and U.S. media have lots of speculation about China's third fourth-generation stealth fighter jet J-18 in addition to the J-20 and J-31 already widely known,

Japan's *Asahi Shimbun* was the first to publish a report on the successful test flight of J-18 at the beginning of 2013. It said that China began to develop its own catapult for the aircraft carrier it planned to build but lacked key technology to make such catapult; therefore, China scrapped its original plan for a carrier for horizontal taking-off of aircrafts, began instead to develop VTOL aircrafts and succeeded in doing so.

Soon afterwards, U.S. *Defense News* weekly published an article that believed that China was developing short-distance vertical taking-off and landing stealth fighter jet, i.e. J-18 Red Eagle VTOL fighter jet, with superb stealth function and installed with laser active phased array radar, internal weapon bays and two vector engines with great thrust.

The Japanese and American reports, though sensational in nature and supplemented by later reports, are but speculation. There had been no evidence on the existence of J-18 whatever until early July, 2014 when Britain's *Jane's Defense Review* published a report on the fighter containing a recent photo copied from a post at a Chinese military forum on the Internet.

Judging by the photo given by *Jane's*, J-18 looks almost the same as J-31 except its canard structure. This gives people the impression that it is a VTOL version of J-31. It sounds reasonable as developing a VTOL version saves money than the development of an aircraft from nothing. This is also the case with U.S. F-35 stealth fighter jet. It has three versions including a VTOL version.

However, J-31 is developed by Shenyang Aircraft Corporation with no experience in developing canard aircrafts. It is suspected that the aircraft is a national project participated by Shenyang and other aircraft manufacturers.

The photo shows that like J-31, it is also powered by two engines perhaps because China lacks of an engine powerful enough to power the fighter alone. But some analysts believe twin engines are safer than a single engine.

However, there is the question: Since China has successfully built an aircraft carrier for horizontal taking off, why shall it develop VTOL stealth fighter jets?

The answer for some people are that there must be a smaller amphibious attack warship carrying VTOL stealth aircrafts to supplement the large nuclear aircraft carriers with electromagnetic catapult that China will build in the future.

I simply do not buy that. The generally accept view is that China will follow the U.S. to be a world police so that it has to have aircraft carrier fleets to send Chinese troops far away to fight battles in other countries.

China shall never do that. The failures of European colonialism and U.S. failure in Vietnam, Iraq and Afghanistan have provided enough lessons that military invasion will bring no benefit to the invader.

European colonialists failed and had to retreat from Southeast Asia in spite of their overwhelming economic and military power, but before 1950s Chinese immigrants without money and without military support from their homeland succeeded there. They now either dominate or are very influential in the countries in that area now.

The above facts teach China that it shall never try to conquer other nations. If it has already had some state-of-art nuclear aircraft carriers, it had better sell them. If it cannot find a buyer for them, it had better scrap them and use the waste steel. Why? Because they are too expensive to maintain.

Then, however, what will China have to defend its trade lifeline in the world?

Amphibious warships with VTOL aircrafts are quite enough, but they are not able to counter the nuclear carrier battle groups of a superpower.

To deal with such groups, China shall develop integrated space and air capabilities, which will enable it not only to defend its trade lifelines but also to counter attack from the space.

Remember, we are now living in the space era!

J-23 and J-25 Stealth Fighter Jets

Israel *Defense Tech* magazine revealed in early January 2014 China's J-23 and J-25 stealth fighter jets that have been already participating in military exercises.

J-23 fighter jet is designed by the Shenyang Aviation Corporation without application of Russian technology. It is based on the F-22 Raptor of the United States to enable Chinese Air Force and Navy to fight a potential conflict with American planes.

J-23 looks very similar to F-22 with its longer body and two V-shaped vertical tails. It depends on the availability of Russian 117S thrust vector engine for improvement of its maneuverability.

J-25, China's one more 4th-generation stealth aircraft is designed by Chengdu Aviation Corporation. J-25 also named "Ghost Bird" is considered one of the best stealth aircraft in the world due to the use of China's 3D printing technology. Analysts say that the main task of the J-25 is to challenge the dominance of the United States and its allies in the airspace over the Pacific, the future environment of Chinese aircraft carriers.

China's Blue Troops

In November 10, 2013, there was a very popular TV series "I Am the Phoenix of Special Force" on show about China's wonderful "Blue Troops".

China now uses its Blue Troops to act as foreign troops in its military drill while Chinese troops are the Red Troops.

Like their foreign counterparts, PLA's Blue Troops adopt tactical principles, organization structure, weapons and equipment as identical to enemy troops' as possible and their soldiers are always in foreign troops' uniform. Some of them even have meals with knife and fork instead of chopsticks to be as closely similar to "enemy" troops.

Being practical, Blue Troops are usually better automatic and high tech than Red Troops just like U.S. troops compared with Chinese troops.

According to Russia media, China's professional Blue Troops are a mystic force often used to carry out special tasks. They are equipped with a J-23 fighter jet that has performance very close to U.S. F-22.

J-23 is a copy of F-22 specially made for Blue Troops. It is a little longer than F-22 with a shape that reduces resistance at supersonic speed. Obviously, it is designed for high-speed attack and interception.

It seems J-25 is developed for Red Troops to counter Blue Troops. If so neither of them are developed for mass production of China's fourth-generation fighter jets.

Mystery of China's 5th-Generation J-60 Fighter Jet Better than U.S. F-22

In early April 2014, a U.S. official told media: China was developing a fifth-generation J-60 fighter jet better than U.S. most advanced F-22 fighter jet in performance. According to him, J-60 is smaller than F-22 but the design of its shape is similar to F-22.

U.S. intelligence community is much worried that China has got details of F-22's design by espionage or Internet espionage. A few years ago Chinese hackers obtained confidential data about U.S. F-35 fighter.

The official said that a short time ago, Chinese media published a photo of the prototype of a new type of aircraft on a truck traveling on Beijing-Shenyang expressway escorted by vehicles of the officials of the Ministry of National Security. U.S. intelligence analysts have made detailed analysis of the photo.

The Chinese media said that the new aircraft was on its way to a stress test center. The aircraft in the photo was a dual-engine aircraft without tail wing or cabin cover. It made some analysts believe it was an L-15 trainer.

However, after analysis of the photo, it was found that the new aircraft was bigger in size than and different from an L-15 trainer in shape of wings and air inlet of engines.

Therefore, it may well be the new J-60 aircraft.

In the past, China made similar official revelation of its new military research project on the Internet. For example, China kept strictly confidential its construction of a new-type Yuan class attack submarine until a photo of the submarine emerged on the Internet in 2004.

Retired USAF Lieutenant General David Deptula who was formerly a deputy chief of staff in charge of intelligence, said that "It should not be surprising that China made public a new type of stealth aircraft."

136

"PLA air force has a set of comprehensive plans and procedures and may have several advanced aircrafts at various stages of design and development," he told a Free Beacon reporter.

There has so far been no additional information about that mystic aircraft.

China's Other Advanced Aircrafts
Mystery of China's Purchase of Russian Su-35s

China's purchase of Russian Su-35 fighter jet has in the main been a done deal. It means that in the period from 2015 to 2020, Chinese air force will have superiority in technology over Japan and India due to its Su-35s.

Su-35 is well known for its maneuverability but it is not a stealth fighter jet. Why shall China import quite a few of it when all focus now is on stealth aircrafts.

According to an article at news.ifeng.com, China's true intention to import Su-35s is not to obtain its 117S engines or Irbis-E passive phased array radar as neither of them is the world or Russia's best. Russia's T-50 fighter will use Type117 engine with much bigger thrust and more advanced SH121 active phased array radar. Therefore, what is really valuable to China in Su-35? It is its K-100 super long-range air-to-air missile with a speed exceeding Mach 3 and a maximum range of 300 km. No air battle weapon in service or being developed has such an unimaginable long range.

Now, there are a lot of early warning, anti-submarine and electronic warfare aircrafts in Japan and the U.S. air force stationed in Asia. The air force system equipped with them has superiority over Chinese system for a long time. China's new generation of fighter jets have almost caught up with U.S. and Japanese ones, but there are still no sufficiently effective means to deal with them as the maximum range of China's air-to-air missiles is only 100 km.

K-100 may be installed with a warhead with active guiding radar or an anti-radiation warhead. The Su-35 equipped with such missiles will enhance Chinese air force's superiority in East Asia. In addition, it will greatly improve the combat effectiveness of China's new generation fighters J-10 and J-20, if it is installed on them.

Some Western military experts believe that China's import of Su-35 reflects its brand new strategic intention to further enhance its existing air and sea superiority in East Asia.

J-16 Fighter/Bomber

In January 2014, Shenyang Aircraft Corp has exposed its new fighter/bomber J-16 and said that it is the most advanced fighter jet in Asia able to contend with Su-35.

J-16 is developed mainly for China's naval air force. It can carry a maximum load of 12 tons of bombs or missiles including YJ-62 and YJ-83 anti-ship missiles. That means that it may gradually replace J-15 for marine battles.

It is more advanced than the fighter jets in service in Vietnam, India and even Japan and is thus able to achieve air supremacy.

Some analysts believe that compared with the Su-30MKK now in service in China's naval air force, there is great improvement in navigation electronic equipment. J-16 is the first fighter equipped with phased array radar, of which the automatic electronic scan is able to deal with multiple targets. It is similar in performance to U.S. F-15E fighter/bomber, but it, in addition, has the capability of C4ISTAR data chain for acting as an AEW&C aircraft so that it can command a small team of fighters in air combat. Moreover, it has electronic warfare capability.

New Version of J-16

Photos of a new version of J-16 appeared at a well-known military forum in late February 2014.

The new prototype J-16 no. 1612 has quite a few improvements compared with prototype no. 1601 that was exposed before. The most significant improvements are the removal of the airspeed tube on its nose and the light gray coating of its radar cover, which indicate that its radar has been replaced by advanced AESA radar. The series number 0102 of its air inlet on its fuselage indicates that small-batch series production has begun and it has been commissioned in PLA air force on trial basis.

As China's J-11BS double-seat fighter is mainly used for air battle and cannot carry lots of air-to-ground and air-to-surface weapons, China has been developing J-16 fighter-bomber that has not only the fine air battle capability like J-11BS but also the ability to carry heavy air raid weapons.

In Zhuhai Airshow 2012, we saw that China had developed quite a few weapons of accurate strike. J-16, as the newest fighter-bomber, will be able to carry all air-launched weapons except strategic ones.

Judging by its photo and recent development in China's aviation industry as a whole, the new J-16 is equipped with AESA radar and HMDS and IRST systems.

It has also adopted certain stealth technology. It is a typical three and a half generation fighter better than the Su-30MKK/Su-30MK2 in service in China.

China's EJ-10B Electronic Warfare Fighter Jet

In May 2014, a foreign media said in its report that China's EJ-10B electronic fighter would be formally commissioned in June 2014. Viewed as a whole, the fighter is a rival to U.S. newest electronic fighter EA-18G.

Chinese air force has been keenly interested in electronic fighters. Its technical staff has often asked about such fighters in international defense exhibitions. EJ-10B is obviously derived from China's fighter jet J-10 with fine performance. We can regard it essentially as a J-10 with quite a few electronic pods. However, there has been little information about it.

It is said that like EA-18G, EJ-10B has made two breakthroughs: an electronic attack aircraft with supersonic speed; and the breakthrough in integrating electronic warfare with conventional warfare.

In a sense, it is an electronic fighter for advanced stage of electronic war. It does not conduct widespread electronic confrontation like the U.S. did during the Gulf War, but provides accurate interference to suppress and destroy enemy target while maintaining electromagnetic dominance as the U.S. did later in Iraq.

The report lists the major equipment U.S. EA-6B electronic fighter has but said nothing about EJ-10B's equipment. It says EJ-10B has made lots of breakthroughs but does not specify.

China's New Mach 2 Stealth Bomber H-18

In October 2013, China made public for the first time its H-6K bomber that carries 6 Changjian-10 (CJ-10) air-to-ground cruise missiles. In addition, H-6K can carry CJ-20 cruise missile with nuclear warhead that China is developing.

However, military analysts believe H-6K is but China's transitional bomber and China is developing a strategic bomber with a range of 7,500 miles.

In March 2014, the website of a well-known U.S. military forum has posted China's plan on H-18 stealth bomber. It is said that H-18 is not a strategic bomber. It has two turbofans with great thrust, a maximum range of 8,000 to 9,000 km and maximum combat range of 3,500 to 3,700 km. Its maximum speed is about Mach 2 and it can carry 12 to 15 tons of ammunition.

It is reported that H-18 has a large internal weapon bay 8 meters long that can carry 72 Leishi-6 small precision-guided bombs or 4 CJ-10A cruise missiles to conduct tactic or nuclear attack against U.S. base at the Guam more than 3,000 km away. It can also carry 4 YJ-12 supersonic anti-ship missiles or 4 YJ-100 long-range anti-ship missiles to kill an aircraft carrier.

It is said that in order to make H-18 better undetectable, it has its air inlet on its back and its wings W-shaped. It has no tail wing and adopts lots of stealth technologies. It will thus be the first supersonic stealth bomber in the world.

China's Newest Medium-range Strategic Bomber B-8 with Rare New Technology in the World

A U.S. military website disclosed in early March 2014 that China was developing on its own a medium- and long-range strategic bomber B-8 with operational radius of 4,000 km without refuel.

Unlike the H-18 described above, there has been no official source about it and it now exists in speculation only.

According to the website, B-8 uses a large integrated navigation/fire control radar capable of long-distance detection and early warning. It has a crew of 4 and can carry two 500,000-ton or one 1,000,000-ton nuclear bomb. It can carry cruise missiles similar to U.S. Pegasus rocket or HN series air-to-ground missiles with a range of 800 or 1,200 km and with a 1,000 to 10,000-ton nuclear warhead or 300-500 kg conventional warhead.

In addition it uses stealth coating, a cold air restraint device to reduce the infrared emission of its engine and other stealth technologies to make it to certain extent stealth. Its internal communications are all carried out by data transmission to be safer and quicker.

There has been no Chinese news about the bomber; therefore, we shall regard the above information as speculation. However, as mentioned in a previous chapter, China is indeed developing medium- and long-range bombers.

China's New J-17 Fighter-bomber Better than Russian Su-34

In late March 2014, a Canadian military magazine reported that China had made new progress in developing its J-17 fighter-bomber based on Russian Su-34 fighter-bomber. What is special is the heavy load of 8 tons J-17 can carry. As a result no Russian or U.S. fighter jet can be its rival in firepower.

China lacks bombers in its air force. It has reached the limit in improving its H-6 bombers; therefore, it has only produced a small number of H-6K bombers, an improved version of H-6 that carries cruise missiles.

China develops a Chinese version of Russia's Su-34 due also to the slow development of its homegrown heavy bombers. That is mainly because China lacks the experience in developing large military transport aircraft.

Chinese version of Su-34 can carry 8 tons ammunition, only a little less than H-6, but is much quicker; therefore, J-17 may be a good replacement for H-6 for China to maintain strong air-to-ground firepower.

In March 2014, the prototype of J-17 had perhaps passed wind tunnel tests. Judging by the model in the wind tunnel tests, it has two seats abreast and a duckbill-shaped radar casing. That means that it can carry one long-range air-to-ground cruise missile at the rack under its belly. It is as good as H-6 bomber.

The report speculated that perhaps China had been deliberating its project of the Chinese-version Su-34 for many years. As far back as in Zhuhai Airshow 1988, there was a footage that displayed the wind tunnel test of a model very similar to Russia's S-34 bomber.

The report said that the codename J-17 is a speculation, but as only the codename J-17 to J-19 have not been used, the new fighter/bombe may be codenamed J-17, J-18 or J-19.

12. Better AEW&C, Radar, Etc. than the U.S.

China's AEW&C Aircrafts

China is a latecomer in the production of AEW&C aircrafts. It first concluded a contract in 1990s to purchase from Israel Phalcon AEW&C but due to pressure from the U.S., Israel terminated the contract. China then turned to Russia for the purchase of Russian ones but cancelled the purchase due to unsatisfactory prices and performance.

However, during the review of military parade in celebration of the 60th anniversary of the founding of the People's Republic of China, China displayed for the first time its homegrown AEW&C aircrafts KJ-2000 and KJ-200.

China's well-known radar expert Wang Xiaomo won China's top scientific and technological award 2012 for his achievements in developing the AEW&C aircrafts. On January 18, 2013, Chinese President Hu Jintao personally presented the award to him.

CCTV reporter Sun Zifa gave a report after interviewing Wang. Sun said in his report that When American experts saw the KJ-2000 and KJ-200 AEW&C aircrafts on October 1, 2009, there was the opinion that Chinese AEW&C aircrafts were more advanced than U.S. E3C as they were the first to use phase resonance radar.

In developing the two types of AEW&C, China got 9 first places in world history of the development of early warning aircrafts, made breakthroughs in tackling more than 100 technological problems and obtained in all about 30 patents. Chinese AEW&C surpass the most advanced mainstream types of AEW&C in the world in quite a few technological indexes such as the furthest range of airborne IT weapon equipment, the largest number of functions and the most complex integration of systems. According to appraisal by American think tank, they are one generation more advanced than U.S. E-3C and E-2C AEW&C aircrafts.

Wang told reporter that in spite of his success in developing independently China's homegrown early warning aircrafts after the U.S. forced Israel to scrap the contract on the provision of an early warning aircraft for China in 1999 and in spite of the top science award he got, he thinks that he has not attained the final goal of his research and there is still a lot of progress to make.

He said, "At present we have to nationalize the entire production in the near future. In the long run, we shall deal with more targets so as to be able to meet the demand of modern warfare. Therefore we have put forth the idea of the third generation of early warning aircraft and the idea has now been accepted by all."

Wang regarded KJ-2000 and KJ-200 as China's second-generation early warning aircrafts as he developed a first generation of unsatisfactory early warning aircraft that was not much used due to unsatisfactory performance. However, KJ-2000 and KJ-200 are regarded by reporters later as China's first generation of early warning aircrafts.

The reporter then asked him, "What is the difference between the current idea and that of the third generation of early warning aircraft?"

Wang said, "We have changed our target. For example, F-22, which is a stealth aircraft. Our goal is to be able to deal with it and to deal with more targets."

During the interview before the award ceremony, Wang explained to the reporter the functions of his early warning aircraft including searching and monitoring targets at sea and in the air and directing China's own aircrafts in fighting. It is the core of modern air combat system that integrates the functions of searching, detecting, collecting intelligence, command and control, telecommunications and navigation guidance, electronic warfare, information transmission, etc.

The report described Wang's dedication to his work. He persisted in working in bed when he was hospitalized for a broken leg in a car accident and for treatment of lymph-gland cancer.

Such dedication is common among Chinese scientists and engineers due to the Chinese intellectuals' mentality reflected in the well-known Chinese saying: "A scholar (Shi) will die for the patron who recognizes his worth; a girl will doll herself up for the man who loves her." Here, the Chinese word of "shi" means not only scholars but also wise self-educated workers and peasants, talented military commanders, kung fu masters, chivalrous swordsmen, etc. It is translated as scholars because scholars are the major part of shi and there is no equivalent term in English.

Since Jiang Zemin's peaceful coup d'état that substitutes intellectuals' dominance of the Party and state for uneducated workers' and peasants' (please refer to Chapter 5 of my book *Tiananmen's Tremendous Achievement Expanded 2nd Edition)*, Chinese leaders know how to recognize intellectuals' worth and value their contributions.

For example, during the Lantern Festival party, Chinese President Hu Jintao

I

was shown on TV screen sitting conspicuously by the side Wang Xiaomo, talking with him. It showed to the whole nation how Hu valued Wang's contributions. Chinese leaders' such attitude of high respect has greatly stimulated intellectuals' enthusiasm in working hard for China's modernization.

Test Flight of China's 2nd-Generation AEW&C

In the preceding section, Wang Xiaomo said in an interview with media after he got the one million yuan (U.S.$160,000) Top Science and Technology Award 2012, that the next thing he would do was to develop an AEW&C capable of detecting F-22 stealth fighter jet.

We believe that as a much respected scientist, Wang would not disclose his ambition if he did not think that it was possible for him to achieve; therefore, it is certainly very interesting. If he has succeeded, F-22 will lose its supremacy.

We are interested whether China's new AEW&C has such a capability. In late May 2013, 28 pictures of what were supposed to be China's new generation of AEW&Cs had been posted on the Internet including ZDK-03 AEW&C for export and a new generation of medium-sized AEW&C for China's own use.

I said in the preceding section KJ-2000 and KJ-200 are regarded as China's first generation AEW&Cs though Wang Xiaomo regarded them as second generation.

It is said that the second generation of AEW&C that Chinese air force and navy will be equipped with, is being test-flied. In a photo related to Wang Xiaomo, father of China's AEW&Cs, we can find a model of China's second generation of AEW&C so covert that we did not discover that mysterious new AEW&C until we had made very careful analysis. (The photo of Wang with three models of AEW&Cs can be viewed at http://tiananmenstremendousachievements.files.wordpress.com/2013/06/wang-xiaomo-and-three-aewc-models.jpg)

In the Photo Wang Xiaomo in his office, there are three models of early warning aircrafts by Wang's side. They respectively are a KJ-200, KJ-2000 and a mysterious AEW&C, which was mistaken as a ZDK-03 for export to Pakistan. However, in fact, a careful observer may find that the coating of the AEW&C model indicates that it belongs to Chinese air force and that its radar is active phased array radar identical to KJ-2000's. As the black edge of the radar is the same as that of KJ-2000's, it uses three-surface solid-state phased array radar. The three black edges of disk-shaped radar case are precisely the wave transmitting

144

material in front of the transmission antenna of the radar.

On the left side is the newest satellite photograph of the second generation of AEW&C developed by China. The photograph indicates that the aircraft's radar is similar to that of KJ-2000's and entirely different from the bar-shaped radar of KJ-200.

Note: There were only photos but no details about the aircraft, not even its codename.

Later in early June, Hubei Daily said in its report at www.cnhubei.com titled "Frequent test flights of China's new KJ-3000 AEW&C equipped with new radar" that recently, some pictures of ZDK-03 AEW&C for export that China had built on Y-9 transport, had been posted on the Internet. There was in addition information that test flights were being carried out of a new generation of AEW&C that would soon be commissioned in Chinese air force and navy. China's homegrown AEW&C again became a hot topic that drew attention both at home and abroad.

It indicated China's efforts in developing an improved version of the KJ-2000 that has already commissioned in Chinese troops. Frequent test flights of the new AEW&C are being carried out. The aircraft that military fans call "KJ-3000" had the number "762" on its fuselage. Compared with KJ-2000, it was more mature in technology and used a new type of phased array radar. Its overall functions were rival to those of E-8, the newest type of AEW&C that the United States has, and it even surpass E-8 in some key technologies."

Soon some more details about KJ-3000 were revealed by another Chinese source.

New AEW&C KJ-3000 Special for Air-to-ground Combat

On June 6, CRI Online, a key news website of China's Central authority run by China Radio International, says in its report titled "Chinese military expert: Improved version of KJ-2000 specialized in directing air-to-ground combat", that The new type of AEW&C disclosed this time was called by military fans 'KJ-3000'. Military experts were of the opinion that the naming of that type of AEW&C was not important. What was important was that the new type of AEW&C had greatly enhanced Chinese air forces and navy's capability of both attack and defense in air combat.

According to CRI Online, Associate Prof. Ge Lide of the National Defense University said, "As far as I know, this new AEW&C uses our homegrown Y-9 transport as carrier. Compared with Y-8 used by KJ-200, there is significant increase in load, range and endurance. At the same time, we see that it uses the large mushroom-shaped antenna of KJ-2000 with long range and relatively strong ability of multi-target identification and tracking. It is entirely capable of both detecting targets and directing and commanding battle. In other words, the AEW&C can not only rapidly detect enemy targets coming to attack it from far away but also distribute targets among Chinese fighter jets and guide them to intercept and hit those targets in the air."

Research fellow Du Wenlong of the Academy of Military Sciences believed that in the future, KJ-3000 would perhaps mainly guide air-to-ground combats and thus provide a better basis for the combat capability of Chinese air force. CRI quoted Du as saying, "Judging by its functions, KJ-2000 is an AEW&C that detects and gives guidance in air combats. If it has to be improved, we are to give further play to its potential and develop a weapon we lack in conducting long-distance reconnoiter and accurate positioning of land targets and giving guidance in attacking them. From this point of view, this KJ-3000 shall be a rival to U.S. E8. Its major function is to give guidance not in air combats but in air-to-ground combats. If our air force is equipped with such AEW&Cs, there will be highly efficient early warning and guidance for our air formation in conducting air-to-ground and air-to-ship attacks. This indicates the commencement of connection between our air force and land combat capabilities and a very satisfactory foundation laid in this respect.'

According to experts, along with the successful trial flights of China's Y-20, in the future, China can develop its AEW&C on the basis of Y-20, which means there will be a larger carrier for China's AEW&C to carry more weapons and equipment to significantly enhance its detecting ability, endurance and directing and guiding ability."

KJ-3000 the Best AEW&C in the World, Much Better than U.S. E-8

On March 18, qianzhan.com carried sino.com report that according to reliable source China has successfully developed its second-generation large AEW&C KJ-3000.

Unlike previous AEW&Cs, KJ-3000 is an integration of all Chinese technology and equipment that are most advanced in the world. Its electronic equipment, are especially advanced, surpassing by far U.S. E8's.

E8 has a rotating round antenna while KJ-3000's antenna is a fixed one that functions universally at 360 degrees. Its multi-layer phased array antenna units are formed in a three-dimensional ellipsoidal array. Network technology is adopted throughout the aircraft. Its new phased array radar is indeed the best in the world. In short, judging by the current financial situation in the U.S., it is very difficult for the U.S. to catch up.

Wang Xiaomo want China's new-generation of AEW&C to be able to detect U.S. F-22 stealth fighter jet. Later in this book, readers will see that China's another new-generation AEW&C KJ-500 can detect F-22. If KJ-3000 can detect F-22, there will be trouble for F-22.

China is developing its Y-20 large transport aircraft with Ukrainian assistance. After successful test flight of a second prototype of Y-20 on December 26, 2013, on April, 2014, there was report that mass production of Y-20 had begun.

If the set of equipment used on KJ-3000 is installed on a Y-20, there will be much room in the Y-20 to carry air-to-air missiles to destroy F-22s if KJ-3000 built on a Y-20 can detect F-22.

KJ-500—China's Another 2nd-generation AEW&C Able to Detect Stealth Targets.

In mid November 2013, a photo of China's new-type homegrown AEW&C has appeared at U.S. Military Forum website and became a hot topic in China and abroad. Military fans have given the aircraft the codename "KJ-500".

In an interview with media, military expert Du Wenlong pointed out: the new AEW&C uses phase array radar with better detection capability than KJ-2000

According to information available so far on the Internet, KJ-500 is made on the platform of Y-9 freight aircraft and is similar in appearance to ZDK-03 AEW&C exported by China to Pakistan.

Web users' speculated that there must be a larger KJ-5000 to form a pair of AEW&Cs similar to the pair of KJ-2000 and KJ-200.

Judging by KJ-500's use disk-shaped phase array radar, it had much smaller blind area and greater distance of reconnaissance. It can track much more targets and direct much more fighters in air combat.

Later in January 2014, a well-known U.S. military forum website published a photo of J-500. According to that website, J-500 must be using a new type of radar with bulging middle. It believed that the new AEW&C has improved capability in detecting stealth targets; therefore, its successful development is of profound and far-reaching significance in China's AEW&C and radar industries. China probably has begun series production of KJ-500.

China's 10 Gaoxin Series Reconnaissance Aircrafts

In late April 2014, huanqiu.com published a report on the Gaoxin series intelligence reconnaissance aircrafts developed by China on its own. Web users have taken photos from unique angles of those aircrafts with strange "cheek" structure when they are flying at low attitude.

Among them, Gaoxin-6 large anti-submarine aircraft is a breakthrough with performance as good as U.S. P-3C anti-submarine aircraft.

Using Y-8 and Y-9 transport aircrafts as platforms, China has developed 10 types of airplanes called Gaoxin aircrafts by changing the fuselages and installing new electronic systems on them for such military and civil purposes as confrontation, electronic reconnaissance, marine patrol, anti-submarine warfare, air control, early warning and air survey. Some of such aircrafts have already been commissioned in Chinese air force and navy.

The report gives description and photos of the various types of Gaoxin aircrafts as follows:

1. Gaoxin-1

Gaoxin-1 is an electronic support reconnaissance aircraft installed with support reconnaissance system and synthetic aperture radar for survey of the battlefield and take accurate image of the land below. It provides direction and guidance with respect to the targets.

Its photo can be viewed at http://tiananmenstremendousachievements.files.wordpress.com/2014/04/gaoxin-1.jpg.

2. Gaoxin-2

Gaoxin-2 is an electronic reconnaissance aircraft developed for the navy. Quite a few antennas have been installed on its fuselage while the antenna case at its nose is larger to hold special large antennas. There is, in addition, a satellite communications antenna on its perpendicular tail wing. It indicates the aircraft's ability to send relevant information to the command post directly through a communications satellite.

Its photo can be viewed at http://tiananmenstremendousachievements.files.wordpress.com/2014/04/gaoxin-2.jpg

3. Gaoxin-3

It is a battlefield command aircraft developed for the air force similar to U.S.

EC-130 equipped with battlefield command and control system. It is mainly used for air-to-air and air-to-land exchange of information and unified command and coordination of various combat units at the battlefield.

Its photo can be viewed at http://tiananmenstremendousachievements.files.wordpress.com/2014/04/gaoxin-3.jpg

4. Gaoxin-4

Gaoxin-4 is an electronic intelligence reconnaissance aircraft developed for the air force. It's special in having large array of antennas by its sides, indicating its ability to capture relatively weak electronic signals. It has relatively strong capability of electronic intelligence reconnaissance.

Its photo can be viewed at http://tiananmenstremendousachievements.files.wordpress.com/2014/04/gaoxin-4-3.jpg

5. Gaoxin-5

Gaoxin-5, also called KJ-200, is an AEW&C. It is entirely produced by China with all China-made components to enable Chinese air force to be entirely independent in having complete compatibility.

Its photo can be viewed at http://tiananmenstremendousachievements.files.wordpress.com/2014/04/gaoxin-5-kj-200-aewc-2.jpg.

6. Gaoxin-6

Gaoxin-6 is a large anti-submarine patrol aircraft as powerful as U.S. P-3. It can carry a crew of 10, including the pilot, radar and sonar operators, and those to search for submarines, make technical analysis, place sonar and operate weapons to form an integrated anti-submarine system of reconnaissance, analysis and attack. It is special in having a very long tail sting, which is said to be China's new MAD as advanced as U.S. P-3C's.

Its photo can be viewed at http://tiananmenstremendousachievements.files.wordpress.com/2014/04/gaoxin-6-anti-submarine-aircraft.jpg.

7. Gaoxin-7

Gaoxin-7 is a psychological warfare aircraft developed by China on its own. It carries the electronic equipment to use standard AM, FM, HF, TV and military communications wave bands to carry out psychological warfare. It blocks the enemy's broadcasts and broadcasts its own programs through those wave bands to the enemy to break the morale of the enemy both military and civilian and achieve the goal of subduing the enemy without fighting.

Its photo can be viewed at http://tiananmenstremendousachievements.files.wordpress.com/2014/04/gaoxin-7-psychological-warfare-aircraft.jpg.

8. Gaoxin-8

Gaoxin-8 is built on the basis of Y-9 transport aircraft with improved engines and propellers. There are enhanced radar on its nose and quite a few antenna cases on its fuselage to collect real-time information about enemy military force and send it to the higher commanders as soon as possible. It is an electronic intelligence aircraft that provides sharp eyes and ears for Chinese navy.

Its photo can be viewed at http://tiananmenstremendousachievements.files.wordpress.com/2014/04/gaoxin-8.jpg.

9. A warning aircraft built on a Y-8 transport

It uses a radar supplied by British company RACAL to provide early warning.

Its photo can be viewed at http://tiananmenstremendousachievements.files.wordpress.com/2014/04/gaoxin-9-early-warning-aircraft.jpg

10. A Y-8 radar testing aircraft

The Y-8 radar testing aircraft tests, adjusts, fine-tunes, confirmatively verifies and appraises the radar under real flying conditions.

Its photo can be viewed at http://tiananmenstremendousachievements.files.wordpress.com/2014/04/y-8-radar-testing-aircraft.jpg

Chinese Radar Detects, Locks on F-22, Causes U.S. Withdrawal of F-22s from Japan

According to the report of an Italian media (the Chinese name of which means progressives) in late March, U.S. military has recently withdrawn its F-22 fighters from Japan back to Guam allegedly for routine overhaul.

However according to informed sources, it is U.S. military and arms dealers' established practice that whenever they have a new weapon, especially a new weapon deployed at an important strategic base to scare the enemy, the weapon must be intensively publicized. It is impossible for the weapon to be withdrawn for routine overhaul.

There must be some mystic reason. According to the Italian media, the major reason for the withdrawal was that an F-22 was discovered and locked on by China's radar that is able to detect stealth fighter jets.

The report says that it has learnt that mystic reason from PLA internal reference material that F-22 stealth fighter was detected and locked on by China's anti-stealth radar.

Does China have such advanced radar?

Anti-stealth Radars, Advanced Electronics Showcased at Beijing Exhibition

The 2014 9th China International Defense Electronics Exhibition opened at China International Exhibition Center (the old exhibition hall at Jingan Village), Beijing on May 8-10, 2014. Lots of strengthened military computers, radar electronic equipment and other defense electronics were displayed there. They drew keen interests from military and electronics fans.

It is the only defense electronics exhibition approved by PLA General Armaments Department. Since its debut in 1998, military media at home and abroad have given intensive publicity to it. In addition, the exhibition provides the perception from high-end professional and authoritative point of view and creative tailor-made services. As a result, it has built up a brand with some influence and has thus further promoted the development of the national defense systems with Chinese characteristics while guiding the integration of military and civil national defense science, technology and industries.

The most attractive exhibits were China's most advanced radars including the following radars that can detect stealth aircraft with detailed description:

JY-50 Passive Radar for Detecting and Tracking Stealth Aircrafts

JY-50 is a new kind of passive system radar. It utilizes the electromagnetic signals transmitted by distributed Radio Frequency stations around it to conduct detection, location and tracking of aerial targets including electromagnetic silence targets. JY-50 radar mainly accommodates air defense warning mission to important direction and sensitive area.

Main functions
- Provide range, azimuth and speed information of aerial targets including electromagnetic silence targets and stealth targets.
- Carry out aerial pre-warning mission to fill up the blind zone in the important direction.
- Support active air information radar to carry out target detection, location

and tracking in the battle direction

Main features
- Passive location system provides strong battlefield survivability.
- Adopts dual/multi-base detection and VHF resonance frequency band techniques to carry out optimum anti-stealth capability.
- High mobility and high reliability.
- Support multi-radar netting, easy to form passive detection network.

Photos of JY-50 radar can be viewed at http://tiananmenstremendousachievements.files.wordpress.com/2014/05/j-50-passive-radar.jpg and http://tiananmenstremendousachievements.files.wordpress.com/2014/05/j-50-passive-radar-able-to-detect-stealth-aircrafts-2.jpg

JYL-1A Radar to Detect and Track with High Precision Stealth Aircrafts and Tactical Ballistic Missiles

JYL-1A is state-of-the-art mobile anti-missile air surveillance radar. It adopts advanced 2D digital active phased array system technology able to provide azimuth, range, height and IFF information of air target (with optional IFF unit). JYL-1A radar accomplishes mid-long range air defense warning mission and guidance mission, meanwhile, carry out detection and tracking mission of tactical ballistic missile.

Main functions
- Aerodynamic target and stealth airplane detection and tracking
- Tracking of multiple tactical ballistic missile with high precision with provision of high-precision tracking data and trajectory estimate data
- Target direction for weapon systems

Two photos of JYL-1A radar can be viewed at http://tiananmenstremendousachievements.files.wordpress.com/2014/05/jyl-1a-long-range-radar.jpg and http://tiananmenstremendousachievements.files.wordpress.com/2014/05/jyl-1a-long-range-radar-2.jpg

The above two radars with capabilities to detect stealth aircrafts are both land-based mobile radars.

JY-27A Air Surveillance & Guidance Radar to Detect, Track Stealth Aircrafts and Missiles

JY-27A is state-of-the-art meter waveband 3D long-range air surveillance radar. It adopts advanced 2D digital active phased array system technology able to provide azimuth, range and height information of air target. The radar has powerful capability to detect stealth target, good anti-jamming capability and mobility. It is featured with excellent airspace coverage performance, high measure accuracy, and excellent track performance for high-speed high maneuvering target.

Main functions
- Conventional aerodynamic target monitor and guidance
- Stealth target detection
- Tactical ballistic missile (TBM) early warning
- Identification of friend or foe (IFF)
- Information analysis and report
- Network data fusion

Two photos of JY-27A radar can be viewed at http://tiananmenstremendousachievements.files.wordpress.com/2014/05/jy-27a-air-surveillance-guidance-radar.jpg and http://tiananmenstremendousachievements.files.wordpress.com/2014/05/jy-27a-air-surveillance-guidance-radar-2.jpg.

This is a long-range land-based radar able to detect and track stealth aircrafts and tactical ballistic missiles.

JY-11B High-Mobility Low-altitude 3D Air Surveillance Radar

The radar operates in S band for tactical air surveillance. It is mainly used for detection of air targets flying at low or very low altitude such as low-flying aircrafts and cruise missiles.

The radar adopts the advanced technological system that incorporates reconfigurable transmitting beam shaping through phase control and Beam Forming Unit for reception. In addition, it widely employs lots of modern radar technologies, such as, low sidelobe planar array antenna, air-cooled solid state transmitter, BFU receiver, DDS-based waveform generator, digital pulse compression, AMTI processing, sidelobe blanking and automatic set-up/tear-down. Therefore, it exhibits excellent operational performance.

Two photos of JY-11B 3D air surveillance radar can be viewed at http://tiananmenstremendousachievements.files.wordpress.com/2014/05/jy-1a-high-mobility-low-attitude-3d-air-surveillance-radar.jpg and http://tiananmenstremendousachievements.files.wordpress.com/2014/05/jy-1a-high-mobility-low-attitude-3d-air-surveillance-radar-2.jpg

There were also other interesting exhibits of electronic warfare in display including the following China's anti-precision strike systems and other equipment and devices of electronic warfare:

Land Spirit Ground-air EO Countermeasure System
The system is an important part of air defense system against the attack of precision-guided weapons. It conducts optical-electronic surveillance and interference to provide protection for important fixed and mobile land targets of enemy precision strike such as the building of national defense, president's residence, headquarters of field operations, air-defense missile troops, bridges, airfields and military strategic points. It can also be a convoy to protect important vehicles and mechanized troops.

A picture showing the operation of the Land Spirit Ground-air EO Countermeasure System can be viewed at http://tiananmenstremendousachievements.files.wordpress.com/2014/05/land-spirit-system.jpg. Two photos of the system can be viewed at http://tiananmenstremendousachievements.files.wordpress.com/2014/05/land-spirit-ground-air-eo-countermeasure-system-2.jpg and http://tiananmenstremendousachievements.files.wordpress.com/2014/05/land-spirit-ground-air-eo-countermeasure-system-3.jpg.

SE-2 Missile Approach Warning System
The system detects UV radiation from the plume of the missile and then sends out alarms to the pilot so that the pilot can take up effective countermeasure to avoid being hit by the missile. The system can greatly improve the survivability of military aircrafts and helicopters in modern war.

There is a picture showing the operation of SE-2 system at http://tiananmenstremendousachievements.files.wordpress.com/2014/05/se-2-missile-approach-warning-system.jpg

DJG8715G Integrated Ground-to-air EW System
The system is an indispensable system for electronic warfare. It can be used to jam the radiation source on an enemy aircraft to protect targets of enemy attack. The system integrates the functions of reconnaissance and interference is therefore an advanced land-based radar jamming system. It can either work alone or be a part of a land-to-air interference network to protect a relatively large area. Moreover, the system may be integrated into an air-defense network to greatly enhance air-defense capabilities. If necessary, it can also jam enemy's radar on land or warship.

Usually, the system is used against:
- Enemy's air-to-ground fire-control radar
- Side view radar
- Missile guiding radar
- Navigation radar
- Morphology tracking radar

Features of the system
- Integrated ESM and ECM functions
- Wide coverage of frequencies

154

- Wide coverage of air battlefield
- Strong ERP
- Strong capability to counter multi-targets from various directions
- Agile and fast management of source of interference

Configuration

- A control vehicle
- An interference vehicle
- A power supply vehicle

Photos and pictures of the system can be viewed at http://tiananmenstremendousachievements.files.wordpress.com/2014/05/djg8715g -integrated-ground-to-air-ew-system.jpg

KG300G Airborne Self-protection Jammer Pod

This pod can be installed on the outside rack of a fighter, fighter-bomber, strike aircraft, medium or small bomber or helicopter to jam enemy I and J wave band weapon radar on aircraft or land.

KG300G adopts the most advanced electronic warfare technology and flexible modular design. Its major combat target is modern radar, especially the pulse Doppler radar carried on an aircraft. The pod can accurately measure the parameters of the target radar. The power management unit in the system distributes the interference resources according to the parameters of the target it has measured so as to be able to jam target radar at the right time, location and frequency.

Product features:

- Adoption of advanced jamming technology
- Adoption of power management and time sharing technology
- High reliability and miniaturization technology
- Satisfactory electromagnetic compatibility with the electronic equipment on the aircraft
- Integrating with the platform carrying it and the equipment on the aircraft
- Able to work independently as well as in coordination with other electronic warfare equipment on the aircraft
- Able to give automatic warning and respond when the aircraft is locked on by radar
- Self examination function
- Highly cost effective

A photo of KG300G Pod together with brief description can be viewed at http://tiananmenstremendousachievements.files.wordpress.com/2014/05/kg300g-ai rborne-self-protection-jammer-pod.jpg

Guangdun Integrated Optical and Electronic Defense System

The system integrates optical, millimeter wave, radar and other means of the surveillance and countermeasure for early warning and accurate location of enemy's low-flying attack aircrafts and missiles so as to destroy enemy's attack capabilities. It counter's enemy aircraft's optical pod, sensors and guiding device for optical guided missile to deprive it of its precision attack capabilities. Its typical feature is the highly effective combination of air surveillance and air search radars and optical-electronic surveillance device to obtain precision location of enemy's attack aircrafts by active and passive means of surveillance.

NRJ5A Shipborne ESM/ECM System

NRJ5A is a new generation of naval electronic warfare (EW) system, which fully meets the operational requirements of modern naval force.

Elaborate design and full utilization of high and new military electronic technologies provide NRJ5A with excellent technical/tactical performance to meet the requirements of future marine wars.

Quick response, multi-target jamming capability and excellent man/machine interaction enable the system to have highly effective jamming capability.

The system can be installed on medium-sized surface combat ships. Due to its modularity, it can be tailored to optimize the capability for different types of ships

Functions:
- Provision of early warning
- Effectively countering target of threat
- EL INT function
- Ability to form integrated combat system with other equipment on the warship

Functions
- Modular design for flexible configuration that can satisfy the requirements of warships of different displacement
- Perfect ability of digital processing and control for quick processing and precise measuring
- Its management of power enables it to optimize its confrontation capability to ensure that the system can effectively jam multi-targets
- It can be connected with infrared and chaff jamming equipment and combined with active and passive interference to achieve effective interference against different threat.

A photo of NBJ5A Shipborne ESM/ECM System can be viewed at http://tiananmenstremendousachievements.files.wordpress.com/2014/05/nbj5a-shipborne-esm-and-ecm-system.jpg

S400 Air Defense Systems from Russia, Putin's Special Favor for Close Ally

British Jane's Defense Review cites Russian Pravda's report on March 31, 2014, as saying that Putin has approved in principle the transfer to China of 4 sets of S400 air defense systems. Alexei Fomin, chief of Russian Bureau of Military Technology Cooperation, said that the two sides have kept on negotiating over the past three years and the negotiation has now reached the final stage.

Previously, there was report that Russian officials always said that unless Russia's own military demand for S400 has been satisfied, there would be no export of it. It seemed the restriction had been relaxed and it might also mean that Russia was now more confident in its ability to protect the intellectual property of its exports.

The report says: Russia has always worried that China aims at copying the system in importing it so that China only imports a small number of the systems.

However, Moscow and Beijing entered into an intellectual property protection agreement in 2008 to enhance protection of intellectual property. In an interview with BBC in 2012, Russian defense analyst Victor Murakhovsky said that China and Russia concluded a new bilateral agreement in 2012, but gave no details of the agreement.

A photo of S-400 air defense missile can be viewed at http://tiananmenstremendousachievements.files.wordpress.com/2014/03/s-400 -air-defense-missiles.jpg

China's Anti-submarine Technology

The Voice of Russia website reported on May 9, 2014 that not long ago a U.S. media published an article pointing out PLA navy's deployment of permanent sonar systems along Chinese coast. U.S. military has also noticed that China is adopting new measures to counter the threat of U.S. and its allies' submarines.

Due to the need to protect China's strategic submarines and aircraft carriers in service and under construction, China's lack of anti-submarine capabilities have been Chinese military leaders' greatest concern. China has to protect such naval assets that it has spent a lot to build.

All China's neighbors have kept on obtaining more submarines. For example, Vietnam, Singapore, Malaysia, Indonesia and Thailand have all been carrying out or drafting their plans to build submarines. Japan has also kept on expanding its submarine fleet while Taiwan has formulated a plan to build 8 submarines with U.S. assistance.

China's goal is to disallow opponents' fleet and air force to enter the waters near China, especially disallowing their free movement within the first island chain. Therefore, establishment of a fixed sonar system in that area is what it can be expected to do. In fact, as far back as in mid 1990s, China has deployed sonar systems at its key ports and naval bases, which people have failed to pay attention to until now. However, the technological performance of those systems has remained a mystery.

Chinese navy obtained some sonar systems from France for its surface warships and submarines in the 1980s and upgraded them later. It has got license for producing some of them or simply copied them. The technological cooperation in that area has been going on in spite of the weapon trade sanction imposed by the West later. In addition, China has got Soviet version of anti-submarine weapons from Russia, Ukraine, Kyrgyzstan and Kazakhstan where there are enterprises producing such Soviet weapons.

While obtaining the technology from the West and the Soviet Union over past decades, Chinese experts have spent lots of resource in their research of submarine technology. However, China's anti-submarine capabilities were quite poor in the 1980s and 1990s when it began to develop such capabilities. As a result, we can imagine that China's fixed sonar systems are a mystery to be unraveled. They may be quite advanced, but may be entirely unable to detect the newest generation of submarines.

However, as it is a vital issue that affects the outcome of any military conflict in the area, obviously, in the foreseeable future, that issue will be foreign intelligence agencies' focus of attention while China will make huge efforts to prevent them from obtaining accurate information about that.

Z-18F Anti-submarine Helicopter, a Powerful New Weapon for China's Huge Warships

On August 12, 2014, China's huanqiu.com, an affiliate of Chinese Communist Party's mouthpiece People's Daily, disclosed China's Z-18F Anti-submarine Helicopter, a Powerful New Weapon for China's Huge Warship.

Z-18F is a new large anti-submarine helicopter for the *Liaoning* aircraft carrier.

People speculate that Z-18 is an improved version of Z-8. It weighs 13 tons and is 23 meters long when its propellers spin. Only the *Liaoning* and Type 071 LPD are big enough to carry it to conduct various anti-submarine, early warning, search and rescue, and transport missions.

Z-8 helicopter was developed on the basis of SA-321 imported from France in the 1970s. It marked China's breakthrough in developing large helicopters when it was displayed at Zhuhai Airshow 2004.

Compared with Z-8, Z-18 has been installed with a new type of engine and blades of synthetic material. Its fuselage has been optimized to improve capability of taking off and landing on high plateau.

As an anti-submarine helicopter, Z-18 can carry more sensors and weapons and have longer range and stay longer in the air due to its bigger size. It has a 32-tube sonobuoy launcher, 7 tubes more than that U.S. SH-60 carries. It can carry four 324mm anti-submarine torpedoes or light anti-ship missiles on its two pylons. Major Western anti-submarine helicopters such as SH-60 and MH-90 can only carry two.

In addition there is a quite big hole at the bottom of its fuselage to let down variable-depth sonar a few hundred meter deep to detect submarines at various depth.

In terms of load, navigation, ability in operating under difficult sea conditions, and anti-submarine detection and attack capabilities, only Europe's EH-101 is a rival to Z-18F, but the former has a better platform than Z-18F.

In addition, due to its big size and loading capacity, Z-18F can do lots of other jobs such as better detection of stealth warships, missile boats, low-flying anti-ship missiles, the air pipe and periscope of a submarine, etc. due to the heavy antenna it can carry.

Z-18F can also provide early warning and direction and midcourse guidance for long-range anti-ship missiles.

It can carry 30 passengers when it conducts rescue mission.

13. Advanced Drones & Transport Aircrafts

China's Excessive Spending on Development of Drones

On June 5, Pentagon said China's military spending last year exceeded the amount of $145 billion published by China.

Pentagon mentioned among other things the advances in Chinese drone technology. It pointed to a Defense Science Board report cautioning Beijing's push "combines unlimited resources with technological awareness that might allow China to match or even outpace U.S. spending on unmanned systems in the future."

It noted that in September 2013, a "probable" Chinese drone was noted for the first time conducting reconnaissance over the East China Sea. China also unveiled details of four drones under development in 2013, including the Lijian, China's first stealth drone, it said.

China's Soar Dragon Better than U.S. Global Eagle in Range and Speed

The website of U.S. "Defense News" weekly published a signed article on January 14, 2013 on China's unmanned aircraft Soar Dragon.

According to the article, Soar Dragon was quite a mystery like a ghost. However, Airshow China has provided some information about that intelligence, surveillance, reconnaissance and information relay drone.

The article gives the following data about Soar Dragon:

Full length: 14.3 m

Wingspan: 25 m

Height: 5.4 m

Normal take-off weight: 7,500 kg

Cruise attitude: 18,000 m

Cruise speed: 750 km/h

Range: 7000 km

According to available information major data of the world well-known high-attitude unmanned reconnoiter aircraft the U.S. RQ-4 Global Hawk are:

Length: 13.5 m

Wingspan: 35.4 m

Height: 4.6 m

Maximum load: 10,400 kg

Power plant: 1 × AE3007H engine, 31.4 Kn (7,050lbf) thrust

Maximum speed: 650 km/h

Service ceiling: 20,000 ft

Endurance: 34 hours

It seems Soar Dragon is somewhat comparable to Global Hawk but with better range and speed.

China's New Lijian Stealth Attack Drone Able to Counter U.S. X-47B

U.S. X-47B stealth attack drone conducted its maiden flight on February 4, 2011 and China's Lijian Unmanned Attack Aircraft conducted, on November 21, 2013. X-47B is regarded as the most advanced of U.S. drones.

On March 21, 2014, qianzhan.com provided some information about the drone in its post titled "China's Lijian drone shocks the world as it can counter U.S. X47B".

Lijian drone is developed jointly by Shenyang Aviation Corp. and Jiangxi Hongdu Aviation Industry Co., Ltd.

According to analysts, it is entirely able to be a rival to U.S. X-47B stealth attack drone in functions and performance and even to challenge U.S. position as the leader in such drones.

It is obviously developed to counter X-47B. X-47B is designed to take off from an aircraft carrier long away from China to deal with the threat of China's anti-aircraft carrier ballistic missile as it can be refueled in the sky. Lijian, being a long-range stealth attack drone, can attack U.S. aircraft carrier long away from China so that there will be no deck for X-47B to take off from.

Y-20 Transport Aircraft

On January 26, 2013, China has conducted a successful maiden flight of Y-20 large multifunction transport aircraft it has developed on its own. It is capable of carrying out long-distance transport of various goods and personnel under complicated weather conditions.

China wants lots of large transport aircraft first of all for transport of troops and military supplies and also for such civil usage as disaster rescue and relief and humanitarian aids. After the maiden flight, relevant tests and trial flight will be carried on according to the plan.

A second prototype of Y-20 was made and had successful test flight on December 26, 2013.

On April, 2014, there was report that mass production of Y-20 has begun.

On February 4, 2014, China's official TV station CCTV disclosed for the first time details of China's Y-20 large transport aircraft, which research and test flight personnel intimately call "fat girl".

The aircraft took about 5 years to develop in order to be able to take off in a simple airfield with a short runway 600 to 700 meters long under poor conditions.

CCTV said that the aircraft had been developed by China solely on its own but both Russian Il-76 and U.S. C17 had been referred to in the process of development. Compared with the Il-76 in service in Chinese air force, there had been lots of improvements in its engines and electronic equipment and increase in it loading capacity.

Details of the aircraft:

Wingspan:	50 meters
Length:	50 meters
Height:	15 meters
Range:	longer than 7,800 km
Maximum speed:	700 km/hour
Service ceiling:	13,000 meters
Power plant:	D-30KP-2 turbofans
Payload:	66 tons
Sweepback angle:	24 degrees

China Is Developing World Largest Amphibious Aircraft

A model of the head of Jiaolong-600, China's new large amphibious aircraft, was disclosed on the Internet in January 2014. It is China's typical new special aviation product. Its test flight is expected at the end of this year.

The aircraft, if made, will be successor of the Shuihong-5 China developed 40 years ago and will be the largest amphibious aircraft in the world, larger than Japan's US-2.

Source says that the large aircraft will satisfy China's need in extinguishing forest fire and emergency rescue at sea. It is installed with 4 WJ6 engines and can carry water tanks on its wings. It is able to take in 12,000 kg water in 20 seconds and travel to and fro between water source and fire site many times. It can keep flying stably at the low altitude of 50 meters and carry 50 people away from danger in rescue operation.

The aircraft is designed and made entirely by China and will have different equipment and facilities for different requirements such as marine environmental surveillance, resources survey, passenger or cargo transportation, etc.

It will be very useful for China in developing its isles and reefs in the South China Sea.

China's 'Caspian Sea Monster'

On April 6, 2014, CCTV carried a high-key report on the successful marine test of China's ground effect vehicle Xiangzhou-1, which is a sea skimmer flying at low attitude above ground and sea with a speed much quicker than a ship but slower than an aircraft. When the first such vehicle developed by the Soviet Union appeared for the first time in the Caspian Sea, it was regarded by people as "Caspian Sea Monster".

Xiangzhou-1 is 12.7 meters long, 11 meters wide and 3.9 meters high with a maximum takeoff weight of 2.5 tons. Its cruise speed ranges between 140 and 160 km/hr. with a maximum speed of 210 km/hr. The speed may be increased to 500 km/hr for a large ground effect vehicle.

As it uses ground effect, it saves fuel compared with an aircraft but is much quicker than a ship. According to international regulations, means of transportation fly at an altitude exceeding 150 meters are regarded as aircrafts but as a boat at an altitude less than 150 meters. In civil application, they shall be certified for their seaworthiness if they are boats. If they are aircrafts they shall be certified for their airworthiness.

China began development of ground effect vehicles in 1967 and has now obtained a series of key technologies with its own intellectual property. It built some small test versions such as DXF100, Albatross and Tianyi-1 ground effect vehicles. In 1999, Tianyi-1 passed examination for certification as a vehicle for inland water transport but was not put into commercial operation for various reasons.

What is China's goal in developing Xiangzhou-1 large ground effect vehicle now?

China plans to use it as a quick means of cargo and passenger transport for its islands in the South China Sea. For example, it takes 15 hours to travel from Sanya, Hainan to Yongxing Island, Xisha (Paracel) Islands by ordinary ship, but less than 2 hours by Xiangzhou-1.

China has lots of islands and reefs in the South China Sea too small for the construction of an airport and too far away for helicopter. Xiangzhou-1 will carry supplies and personnel in relatively poor weather to the islands and reefs within a short time and at low costs.

That proves that China has made a plan to develop its islands and reefs for quite a long time; therefore, it has to first develop the ground effect vehicle for transport. When the means of transport is available, it began large-scale reclamation on five reefs and oil exploration there. The artificial islands built on reefs by reclamation will be China military and fish farming bases. Now with Xiangzhou-1 ground effect vehicles China can send goods, personnel and supplies to the islands and oil rigs for the livelihood of the personnel and the maintenance of the equipment there.

Ships are too slow, helicopters are unable to travel such long distance and aircrafts lack airport; therefore, ground effective vehicle is the only suitable means of transport for China's large-scale oil exploitation and development of fish farming and tourism in the South China Sea.

14. Missiles, World Fastest Helicopters, and Others

China's DF-15C Bunker-Busting Penetrator

In October 2013, China's *Ordnance Knowledge magazine* discloses for the first time China's DF-15C, a new version of DF-15 missile designed as a powerful bunker-busting penetrator.

The missile has a range of about 700 km and accuracy between 15 and 20 meters due to its use of radar guidance or infrared imaging guidance at terminal stage. It has a warhead 2 to 2.5 meters long, the biggest in Chinese missiles. According to description, the warhead is mainly used to penetrate enemy bunker, reinforced concrete strengthened permanent fortification, underground facilities and command center.

Four photos of the missile and its explosion and launch can be viewed at http://tiananmenstremendousachievements.files.wordpress.com/2013/10/df-15c-missile-1.jpg, http://tiananmenstremendousachievements.files.wordpress.com/2013/10/a-group-of-six-df-15c-missiles.jpg, http://tiananmenstremendousachievements.files.wordpress.com/2013/10/explosion-of-df-15cs-warhead.jpg (explosion of DF-21C warhead), and http://tiananmenstremendousachievements.files.wordpress.com/2013/10/e59bbee78987e4b8bae69c80e697a9e69b9de58589e79a84df-15ce58f91e5b084e785a7e78987e38082.jpg

China's Terminal-Sensitive Projectile Electronic Weapons

In December 2012, chinanews.com published a report on the successful test of China's terminal-sensitive projectile ammunition, a kind of electronic weapons with artificial intelligence. It is a kind of most advanced weapon that China has succeeded in developing through 2 decades of efforts. A photo of its successful test can be viewed at http://tiananmenstremendousachievements.wordpress.com/2012/12/06/test-of-chinas-terminal-sensitive-projectile/tank-destroyed-from-above-by-terminal-sensitive-projectile/

In the photo, there are tanks destroyed from above by a terminal-sensitive projectile.

Terminal-sensitive projectile is a kind of cluster bomb containing a large number of warheads with artificial intelligence able to detect the targets and turn the warhead toward them when it explodes.

It is now the most advanced long-range anti-tank weapon characterized by its large attack range, high accuracy and outstanding damage effect. When used, it can destroy a fleet of tanks a long distance away.

In addition, due to its simple structure, it is very cost effective.

According to Yang Shaoqing, chief designer of China's terminal-sensitive projectiles, the U.S. spent U.S.$1.7 billion in more than 2 decades in developing its SADARM terminal-sensitive projectile. Germany spent U.S.$600 million in 12 years and so did Russia. China spent only a fraction of what developed countries have incurred in developing its own terminal-sensitive projectiles in 2 decades.

Crab torpedo and Super Hornet Program to Destroy Aircraft Carrier and Its Aircrafts

China has developed crab torpedoes and super hornet program to destroy U.S. aircraft carriers

In late February 2014, boxun.com quoted New York Times as pointing out that China had been going all out in carrying out its ocean killer plans mainly aimed at establishing an independent "Explosive Crab" combat system at sea and a "Super Hornet Program" for superiority in sea battle. Such plans will severely challenge U.S. hegemony at sea.

According to Pentagon intelligence, an "Explosive Crab" is a wake homing torpedo with improved design. When it leaves a warship or submarine, it goes 100 miles away on its own and lies waiting there until the time it is instructed to rise to the surface to receive the newest instruction from a satellite to launch a sudden lethal attack at an enemy ship.

In war such deep-sea killers can be deployed by submarines along major or possible cruise routes of enemy warships and aircraft carriers to launch surprise attack at U.S. Navy.

The United States admits that so far there is almost no adequate defense against such torpedoes. It is very much possible that if there is a war between China and the U.S., lots of "Explosive Crabs" will be deployed along U.S. coast to blockade U.S. marine shipping and trade.

In addition, China has been developing a "Super Hornet Program" that scares the U.S. It is a group of tiny drones with artificial intelligence designed to destroy all the aircrafts on the deck of an aircraft carrier.

China to Be World Biggest Missile Producer

According to U.S. Aviation Week & Space Technology's report in early January 2014, China is growing into the biggest missile manufacturer in the world. Two of China's major arms manufacturers will turn out 50,000 missiles with terrible firepower.

Previously, there was a report that China had deployed 1,000 missiles targeting Japan for victory in its confrontation with Japan. That number is but a fraction of China's missile production capacity. Such a huge number of missiles may enable China to subdue its enemy without fighting.

According to the data provided by U.S. analysts, in the coming 5 years, China will be the leader on world missile market. Its Norinco will rank the first and is expected to produce 29,992 missiles, taking a 15% world market share while U.S. Raytheon Company will rank the second and is expected to produce 23,744 missiles with a world market share of 12%.

China's CPMIEC will rank the third with expected production volume of 18,479 missiles and a world market share of 9%. Europe's MBDA, the fourth with production volume of 11.232 missiles and market share of 6% while Russia's Tura arms plant, the fifth with production volume of 10,413 missiles and market share of 5%. All the other missile producers are expected to produce a total of 108,787 missiles with a market share of 53%.

In the fiscal year of 2014, in terms of the value of products, U.S. Raytheon is expected to turn out USD10.1 billion worth of missiles with a market share of 15%, ranking the first; U.S. Lockheed-Martin, USD5.8 billion with a market share of 9%, ranking the second; while China's CASC, USD5.3 billion with a market share of 8%, ranking the third.

China Has Made Huge Investment in Lots of Advanced Cruise Missiles

U.S. The National Interest website published an article on May 12 titled "China's Cruise Missiles: Flying Fast Under the Public's Radar", stating China has deployed a large number of increasingly advanced cruise missiles, but compared with its ballistic missiles, they have roused far less concerns. However, their impact on regional security, deterrence and potential military operations may be as great as ballistic missiles. In fact, the lack of public concerns is precisely why cruise missiles are so attractive to China.

The authors of the article say, "China's military modernization is focused on building modern ground, naval, air, and missile forces capable of fighting and winning local wars under 'informatized conditions.'" For this, China has to hinder or defeat U.S. intervention by its anti-access/area denial (A2/AD) capabilities. They believe that a large number of highly accurate anti-warship and land-attack cruise missiles are an important element of such capabilities. Sources say that Guam and Australia's Darwin are now within the range of China's cruise missiles. China has developed its own high-performance anti-ship cruise missiles and imported supersonic anti-ship cruise missiles from Russia. It now can launch cruise missiles from land, aircrafts, warships and submarines. In fact, all China's new surface warships and conventional submarines are able to do so.

The article says China is faced with some challenges in fully exploiting its increasingly large arsenal of cruise missiles. The three major challenges are:

First, does China have the C4ISR to make the best use of its anti-ship cruise missiles? China shall be able to find the locations of its mobile targets and communicate the information promptly to the units that are assigned the task to launch the cruise missiles to attack the targets.

Second, can China carefully organize and direct with good coordination a complex, multifaceted air and missile campaign for many days? The success of such a campaign depends on both human and technical factors. In addition to obtaining the advanced cruise missiles, China has to satisfactorily train necessary military personnel to enable them to get years of experience in handling such campaigns in various ways and commanding the control system of such combine-arms operations. So far, it is unclear whether China has such ability.

Third, people generally believe that the revolution in information technology and progress in satellite technology have made it easy for cruise missiles to hit their targets accurately. However, skillful application of the technologies is not as simple as possessing the technologies.

China has made huge investment in obtaining foreign cruise-missile technology and developing its own products. Such investment has enabled China to deploy advanced anti-ship and land-attack cruise missiles on its new and old warships, submarines, aircrafts and land force to enhance PLA's combat effectiveness. The cruise missiles are important elements of the A2/AD capabilities China has been building up.

China's New YJ-91 Anti-radiation Aegis Destroyer Killer Missile

Recently a certain air force division in Shenyang conducted a drill with real arms and equipment for reorganization. In some official photos published by PLA, a Flying Leopard fighter-bomber is seen carrying a new type of missile under its wing. Judging by its appearance, the missile is an YJ-91 missile that has recently been exposed. It looks like a Russian KH-31P anti-radiation missile.

In 2012, Voice of Russia published a report, stating that in late 1990s, Russia exported KH-31P anti-radiation missiles to China. Soon afterwards, China imported the technology and was licensed to produce such missiles. It gave the missile the codename YJ-91. Before that, China tried to produce an anti-radiation missile on its own but failed.

The new anti-radiation missile produced with licensed Russian technology becomes Chinese air force's major weapon in breaking through enemy air defense. China has specially developed special fighter jets to carry it, for example, China's J-8HG fighter-bomber.

China's WZ-10 Armed Helicopter Better than AH-64D Apache

According to a report by www.news.com on November 18, 2012, five days earlier, China's indigenously developed WZ-10 armed helicopter made its debut flight demonstration at Zhuhai Airshow.

Ten photos of the helicopter can be viewed at:

http://big5.xinhuanet.com/gate/big5/news.xinhuanet.com/photo/2012-11/18/c_123966765.htm

The report says that according to China Central TV (CCTV), recently a live ammunition drill was carried out by Nanjing Military Command in East Anhui. WZ-10 took part in the drill, the first time ever for the new armed helicopter.

It displayed its air-to-ground and air-to-air attacking ability with super maneuverability and creeping ability. In particular, it has much greater firepower as it is able to carry much heavier load.

The report says that the helicopter means a great leap forward in Chinese army's air attack power.

Hong Kong's Singtao Daily says that its reporter watched a 20-minute flight display of the helicopter at close range at Zhuhai Airshow. He was profoundly impressed by its super low flight at high speed, reverse flight, perpendicular climbing up and down and low noise. He said that he could barely hear the noise of its engine.

Singtao says according to Wu Ximing, the chief designer of the helicopter, compared with U.S. AH-64D Apache Longbow, WZ-10 is a little inferior in its engine, load and fire power, but is far superior in maneuverability and air combat capability.

China Has Developed Powerful Engine for New Heavy Armed Helicopter

China's aviation industry has long been weak in the production of helicopters. However, China has great demand for powerful heavy helicopters.

Its navy needs anti-submarine, transport and rescue helicopters on its aircraft carriers, destroyers and frigates and attack helicopters for its landing platform docks.

Its army needs heavy attack and transport helicopters, especially those that can function well on high plateau where armed conflict remains possible between China and India.

Previously China imported some S-70 Black Hawk helicopters from the United States but had to import helicopters from Russia when Black Hawks were not available. The Russian ones proved inferior to S-70 especially at high plateau.

Through comparison, China finds that Black Hawk's fine performance is due to its powerful engine. China has been particularly weak in aviation engine. It cannot give full play to WZ-10's wonderful functions also due to its inability to get powerful engines from Canada or make the engines itself. Therefore, it has been making great efforts to catch up quickly.

Now, according to huanqiu.com's report in mid April, 2014, China has successfully developed jointly with a French company its Wozhou-16 engine as good as the T700-GE-700 turboshaft engine with about 1,200 to 1,500 kW power greater than the 1,210 kW of T700-GE-700. The engine is entirely digital electronically controlled and is efficient in fuel consumption. In addition, it can operate for more than 3,000 hours before it needs an overhaul.

For WZ-10, China has developed 1,500kW Wozhou-15 and resolved its engine problems. As a result, it has improved its engine, load and firepower to be better rival of AH-64D.

With Wozhou-16 engine, China developed its Z-20 all-purpose helicopters to satisfy the needs of its navy and army.

China Successful Maiden Flight of Z-20 New Type 10-Ton Grade Helicopter

With its success in developing a powerful engine as mentioned in the preceding section, at 1120 am, December 23, 2013, China successful conducted maiden flight of its new-type Z-20 helicopter at a certain airport in Northeast China.

170

China has developed its Z-10 and Z-19 military helicopters but lacks a medium-sized 10-ton grade one. The new Z-20 is a multi-purpose one. It can be used for reconnaissance, air strike, transport and logistic support.

It looks like U.S. Black Hawk helicopter, but with a different five-blade propeller that makes it easy to control and improves its maneuverability. The transitional structure between its body and tail optimizes its transport performance.

It is expected that the helicopter will be deployed on China's amphibious attack warships in dealing with the disputes over islands and reefs.

Drill of Lots of WZ-10 Apaches Reflects Intensity of China's Military Buildup

Chinese military's official website 81.cn reported in mid March 2014, a certain regiment conducted a military drill according the requirements of real war including training in such risky and difficult operations as tactical formation with various equipment, hedgehopping and fire assaults in order to enhance the troops' all-weather combat capability.

China is now producing quite a few types of helicopters. China began conducted series production of WZ-10 armed helicopters in 2009 and delivered them to the PLA. The scale of the drill indicated the intensity of Chinese military buildup. In less than 5 years, China has built and trained so many advanced helicopters and pilots to obtain substantial anti-tank capabilities.

A photo of the drill of lots of WZ-10s can be viewed at tiananmenstremendousachievements.files.wordpress.com/2014/03/wz-10s-in-milit ary-drill-23jpg

China's Z-19 Learns from U.S. by Being Equipped with Millimeter Wave Radar

According to the photos recently posted on the Internet by a military forum, like U.S. AH-64C/D Apache Longbow and Russian Mi-28N Night Hunter, China's Z-19 armed helicopter has also been equipped with a millimeter wave radar, which looks very similar to the fire control radar on U.S. AN/APG-78 Longbow. Therefore, some U.S. military forums complained that China had copied U.S. weapons again.

Sources say that PLA's Z-10 will also be upgraded by the installation of millimeter wave radar. As a result, China will be the third country in the world to have equipped its armed helicopters with millimeter wave radar.

Due to the delivery of AH-64E armed helicopters to Taiwanese army, the PLA now pays great attention to improvement of its Z-10 and Z-19 helicopters. One of the approaches for the improvement is the installation of millimeter wave radar on them to greatly improve their detection ability and fire control.

WZ-10 has been developed due to the emergence of Apache. However, China's ambition in developing helicopters does not stop there. I am going to describe below an unmanned stealth and a super fast helicopter that China is developing with new ideas.

Test Flight of Proof Version of China's New Unmanned Stealth Helicopter Jueying

China's new-type stealth unmanned helicopter Jueying with coaxial contrarotating propeller was disclosed for the first time at China Helicopter Exposition in Tianjin. In late September 2013, new information was revealed again at the unmanned aviation master competition in Beijing.

Now research, development and test of Jueying unmanned helicopter goes on. The key technical problem to be resolved is the air disturbance to the lifting propeller caused by the airstream generated by the high-speed forward propeller. The working staff said that when the forward propeller operated at high speed, it generated a huge back rotating airstream, which greatly affected the distribution of airstream under the lifting propeller.

The reporter has learnt that at present the project has entered the phase of test flight to prove through the test flight of that version the influence of the various airstreams on the helicopter when it is flying.

The final version of Jueying will be 2.8 to 3 meters long and fueled by oil and gas. It has been decided that shape and coating stealth technologies will both be used. It will be an unmanned helicopter with both reconnoiter and strike functions.

Judging by the footage displayed at the Exposition, all the landing and taking-off gears can be taken in to reduce air resistance. The target speed of design exceeds 300 km/hour, which ranks first among small unmanned helicopters.

A photo of the model of the unmanned stealth helicopter can be viewed at http://tiananmenstremendousachievements.files.wordpress.com/2013/09/img-4248.jpg

China Developing World Fastest Helicopter with Speed Exceeding 700km/h

On September 13, 2013, SCMP reported that Chinese engineers were working hard to develop a super fast helicopter codenamed Blue Whale with the most innovative and risky designs on drawing board.

Having successfully applied technology learnt from other countries, they were now trying to break new ground in developing the fastest helicopters in the world.

One of their ambitious designs was a helicopter with four rotors called "Blue Whale". It is a project by the China Helicopter Research and Development Institute. The four rotors in Blue Whale can tilt from a horizontal to vertical position and vice versa. They will enable the helicopter to reach the speed of more than 700km/h, 40% faster than Boeing's V-22 Osprey.

SCMP says, "With a cruising speed of 538km/h, developers say the Blue Whale will lift 20 tons of cargo and fly more than 3,100 kilometers without refueling. It will fly as high as 8,600 meters, also higher than the V-22."

The institute's director Qiu Guangrong said last year that the institute hoped to build a high-speed helicopter within 5 years without specifying what model it would develop.

At that time, 200 km/h seemed the speed limit for most helicopters as the huge stresses on engines and mechanical parts at high speed may cause them to fail or break apart.

There have been lots of attempts for breaking the speed limit with limited success. Boeing has developed a twin-rotor design with one rotor spinning in opposite direction to the other to avoid turbulence of air, but neither Boeing's model nor Europe's experimental Eurocopter X3 with a conventional rotor plus twin propelling rotors on fixed wings achieve a speed exceeding 500km/h.

Blue Whale will have 4 rotor pods mounted on fixed wings to provide lift when they are in the horizontal position. When the helicopter is airborne, the rotors that have turned vertical provide forward thrust.

The government has allocated huge funds to the project due to PLA's need for fast helicopters.

There is some doubt whether a functioning prototype of the Blue Whale may be made within five year. The unique problems with the four-rotor design are difficult to overcome. The trickiest one is how to compensate for the turbulence generated by front two rotors on the rear two during flight.

SCMP said in its report, "Some in the industry also doubt whether China even needs to develop high-speed helicopters while it is still struggling to catch up with developed nations in conventional helicopters.

"The weakness of domestic helicopters was exposed during rescue efforts after the Sichuan earthquake in 2008. The PLA sent its most advanced helicopters to the area, but television footage showed that the bulk of the rescue work was carried out by Black Hawk helicopters imported from the U.S. in the 1980s."

However, it is 2014 not 2008 now. In a previous section, I mentioned that Chinese experts found Black Hawk's wonderful performance was due to its powerful engine. In mid April, 2014, China has successfully developed jointly with a French company its Wozhou-16 engine as good as the T700-GE-700 turboshaft engine with about 1,200 to 1,500 kW power greater than the 1,210 kW of T700-GE-700. Moreover, China has produced engines with 1,500kW power for its WZ-10 armed helicopter, much greater than the 1,300kW of the Canadian engine that WZ-10 was initially designed for but could not get from Canada due to U.S. interference.

The above-mentioned high altitude is no longer a barrier for the helicopters China is producing now.

Like China's HGV, aerospaceplane, ASAT, anti-ASAT and ICBM interception projects, the Blue Whale project signals the beginning of a new era when China no longer follows other countries' steps to catch up with them but is developing what other countries have not tried to do or have been trying without success.

As mentioned before, people outside China, especially Americans shall be on their strict guard not to underestimate China's potential.

PART 3   THE NEW COLD WAR

15. China's Anti-Encirclement Diplomacy

Unrequited Love

American people are so happy at Obama's words that the United States will never be the number two in the world. That is perhaps a part of the American Dream at present.

For Chinese leaders and people, however, the Chinese dream is but to grow powerful enough to put an end to China's miserable past of being bullied by foreign powers.

As China has not grown powerful enough yet, China relied on the United States to avoid being bullied.

As mentioned at the beginning of Chapter 6, like a poor girl courting a handsome and wealthy man, China has done almost everything to please the man.

First, the U.S. wants a peaceful solution of the Taiwan issue. China has done the best to win over Taiwan peacefully.

Second, China avoids persecution of the dissidents the U.S. openly supports and even allows them to leave for the U.S.

You may protest: What about Liu Xiaobo?

Liu Xiaobo is imprisoned for advocating multi-party democracy, i.e. depriving the CCP of its monopoly of state power. In my book *Tiananmen's Tremendous Achievements Expanded 2nd Edition*, I describe the current Chinese regime as the CCP Dynasty. Does any dynasty in Chinese history allow its monopoly of state power to be deprived?

Third, China is willing to incur heavy losses to please the United States.

The most obvious example was China's support of American actions in overthrowing Libyan dictator Gaddafi in spite of the heavy losses China might and has indeed suffered due to that.

The U.S., however, always looks down on China due to China's one-party autocracy, poor rule of law and human rights and keeps on neglecting China's courting.

Encirclement of China

As Chinese economy continues its tremendous growth and has surpassed Japan, the U.S. began to regard China as a real threat that may replace it as the number one in the world.

In fact, to remain number one, the U.S. has always been restraining its potential contenders such as Russia and China, what is different now is that the U.S. has begun to encircle China.

The U.S. calls its new strategy of encircling China its "return to Asia".

What does it mean? In fact, there is no need to return as the U.S. has never left Asia. However, Asian countries that have territorial disputes with China, immediately realized that what the U.S. meant was that the U.S. would be the big brother in Asia to support them in their disputes with China.

Before U.S. "return", there were disputes but no tension. Soon after U.S. declaration of its "return", with the big brother behind them, all contenders have risen up to challenge China. China soon found itself tightly encircled.

However, the parts of the encirclement to the north and west of China are Russia and India and Vietnam who relied on Russian support in their territorial disputes with China.

When the U.S. was restraining China, Russia was doing so too as it thought that China's rise might be a threat to it; therefore, Russia kept on supplying India and Vietnam with weapons a little better than those it supplied China.

The U.S. certainly wants to play its role in having India and Vietnam joining U.S. encirclement of China. When Leon Panetta was U.S. defense secretary, he visited the two countries in 2012 in addition to other Asian countries for that purpose.

Encirclement of Russia, a Byproduct of Encirclement of China

Like German strategists in World War II, U.S. strategists miscalculated.

Germany knew well that it should not fight a war on both eastern and western sides, but it attacked the Soviet Union and finally lost the war for that.

Now, in formulating the new strategy of "return to Asia", the United States neglects the potential of Russia, the major part of the disintegrated Soviet Union with the ambition to succeed the Soviet Union as another number one in the world, i.e. an equal rival to the United States.

At that time, Russia was in dire predicament. Almost all former Soviet constituent and satellite countries have been or are being drawn away by the EU. Russia had almost no influence in Europe.

In the strategic areas for its security in the Middle East, the West led by the United States have been cutting Russian ties with the countries there, first Iraq, second Libya and then Syria. It seemed that Russian influence would remain only in Asia mainly in India and Vietnam.

178

If by its return to Asia, the U.S. has substituted its influence for Russia's in India and Vietnam and subdued China by encircling it, then it will be Russia's turn to be closely encircled by the U.S.

China Broke the Encirclement by Allying with Russia

Immediately after U.S. change in strategy, China altered its strategy too. It was no longer a faithful follower of the United States. It switched to Russian side by the use of its veto to help Russia keep its influence in Syria. That marked the commencement of a Cold War alliance between China and Russia in countering the U.S. though at that time perhaps China merely wanted to break the encirclement on its north and east.

While U.S. Defense Secretary was busy in Asia to draw India and Vietnam to U.S. side so as to complete U.S. encirclement of China, Russia used its influence for improvement of Chinese relations with India and Vietnam. As a result, Chinese Premier Li Keqiang was able to successfully improve China's relations with India in his visit to India in May 2013. Vietnamese President Truong Tan Sang visited China in June 2013 to improve Vietnam's relations with China.

The western and northern parts of the encirclement have been broken.

The breakthrough in easing tension between China and Vietnam has indicated China's willingness to cooperate with South China Sea claimants in fishing and extracting oil and gas in the disputed waters. As a result, both Malaysia and Brunei refrained from joining the encirclement.

Due to its need for China to restrain North Korea, South Korea has refrained from joining the U.S. encirclement.

Taiwan has been building economic ties with Mainland China. Even in the pro-independence Democratic Progressive Party, there are lots of people who want to have a "one-China consensus" with Mainland China.

Now, there are only Japan and the Philippines that remain U.S. allies in encircling China, not enough even to encircle China merely on China's east side.

Putin Visited China for Much Closer Ties in June 2012

China's veto on Syria issue pleased Russia so much that Russian President Vladimir Putin wanted closer ties with China.

He chose Beijing as the first destination of his visit abroad after he was reelected Russian president. He made the choice to show the importance he attached to relations with China. Before his visit in June 2012, he published an article in China's official mouthpiece People's Daily, stressing that Sino-Russian cooperation was not directed at any third Party, but pointing out that without Chinese and Russian involvement, no international issue could be discussed or enacted. He wanted much closer partnership with China in economy, trade, energy, culture and technology.

In his meeting with Chinese President Hu Jintao, they signed a joint declaration for further development of the two countries' strategic partnership. The two presidents also pledged to intensify military exchanges, strengthen co-ordination in Asia-Pacific and co-operate in investments.

After their meeting the two presidents attended the signing of 11 cooperation documents, including cooperation between the two countries in building Nos. 3 and 4 generating sets in Jiangsu's Sutian Bay nuclear power plant and the establishment of a U.S.$4 billion joint investment fund.

Putin brought with him lots of Russian ministers and most major business executives. The two countries would carry out intensive discussion for a long time to enhance cooperation. This seemed to be a second honeymoon between the two countries after their first honeymoon when China pleased Stalin by sending troops into Korea in the 1950s.

With China and Russia as the centers, the six-country Shanghai Cooperative Organization (SCO) would adopt its first comprehensive plan to expand the bloc from focusing on security co-operations to being an economic and geopolitical alliance. Moreover, SCO planned to enlarge the alliance. It had already taken in Iran and other three as observer states and planed at that time to take in Afghanistan and turkey respectively as observer state and dialogue partner at its June summit meeting.

Putin had announced his ambitious plan to build a powerful navy while China was vigorously building up its navy silently. The U.S., though lacked money, planned to increase the proportion of its navy in the Pacific.

The arms race had already begun and the mutual blocking was actively underway! The U.S. could not do what it wanted in Syria without Chinese and Russian involvement while China could not smoothly settle its South China Sea disputes due to U.S. support to other claimants.

It seemed Putin had started a new Cold War.

Xi Jinping Failed to Win Putin's Support in China's Standoff with Japan

However, in spite of the promise to intensify bilateral relations and the plan of the Shanghai Cooperative Organization to expand the bloc from focusing on security co-operation to forming an economic and geopolitical alliance, not much progress has been made in strengthening the alliance in the year since then.

However, the situation was quite different later in early 2013. There was possibility of real military confrontation between China and Japan and U.S. involvement in the confrontation. China was in dire need of Russian advanced weapons and technology.

Like Putin, in order to show the importance he attached to Sino-Russian relations, Xi Jinping chose Russia as the first destination of his foreign visit since he became Chinese president.

Russia was still bothered by the situation in Syria and Iran as its influence there was quite important.

In addition, both China and Russia were unhappy at Western criticism of the human rights situation in them; therefore, to please Russia and also for China's own sake, Xi said in his speech at a Russian international relations school, "We must respect the right of each country in the world to independently choose its path of development and oppose interference in the internal affairs of other countries."

The U.S. was certainly afraid that the close partnership between China and Russian might grow into an alliance against the West and initiate a new Cold War against the U.S.

There was news at that time that U.S. decided to suspend its plan to deploy its anti-ballistic-missile in Europe, which Russia had opposed strongly for a long time. That was certainly an effort to win over Russia.

China did not share Russia's worry in that respect, but its successful test in intercepting an ICBM before Xi's visit certainly impressed Russia.

China and Russia both had maritime territorial dispute with Japan. Xi wanted to draw Russia to China's side so that he said in the above-mentioned speech that China and Russia should jointly preserve the achievements of the victory of World War II and the international order after the War.

Xi meant that as Russia got the four islands in dispute with Japan due to World War II, China supported Russia in the dispute. On the other hand, as Japan returned Taiwan to China due to its defeat in World War II, it should give back to China the Diaoyus (known as Senkakus in Japan) as the Diaoyus were a part of Taiwan while Taiwan was a part of China.

Xi certainly wanted closer military cooperation with Russia and if possible,

Russian support in China's dispute with Japan.

However, Russia did not respond to Xi's words, but allowed Xi to be the first foreign leader to visit the Armed Force Action Administration Center of Russian Defense Ministry. Xi said that the visit of such an important place of Russian military indicated the intensification of military, political and strategic cooperation between Chinese and Russian armed forces.

However, the visit may be symbolic. What people were interested in was whether there would be any breakthrough in Sino-Russian cooperation in weapon development.

For example, both China and Russia wanted to build nuclear aircraft carrier. Russia was strong in developing the nuclear reactor, while China had succeeded in developing electromagnetic catapult. Should they not each provide the other what the other needs?

In fact refusing to provide did not work. Russia refused to provide arresting cables to China's first aircraft carrier, but China succeeded in making them on its own. Supplying what the other lacks is a way to recover research costs and make a substantial profit. Both China and Russia must be clear that the other will after all get what it wants through the efforts of its own research workers and engineers.

It is the same with the U.S. The U.S. does not allow China to join the international space station, but China has now developed its own. What if China stores star war weapons in the bigger space station it plans to build? The U.S. has just lost control.

Close cooperation between China and Russia in weapon development will have the consequence of adding wings to tigers for both countries. That will be what the U.S. fears most. That did not seem to happen soon at that time as it took time to establish mutual trust and maybe there would never be the mutual trust required.

However, at that time there was the possibility of such cooperation and even the emergence of a real alliance between China and Russia.

Moreover, if there was such a sound alliance, a war between China and Japan would be much more likely. Due to Japan's cruel occupation of China for 14 years in the past, there is inveterate hatred against Japan among quite many Chinese people though in history China was not a nation fond of retaliation. However, if Japan persists in denying its error and justifying the war, it is quite hard to avoid a war when Japan tries to amend its constitution for military buildup.

The close alliance between China and Russia might enable China to win the war and prevent U.S. involvement as being so remote from Asia, U.S. was entirely unable to fight both Russia and China in an all-round war in Asia.

For Xi Jinping, the victory of a limited war with Japan would enable him to consolidate his power and in addition provide Chinese troops the opportunities to test their new weapons and obtain actual war experience.

There was the economic side of alliance that would be mutually beneficial and much more possible.

As the EU had attracted almost all the satellite and constituent states of former Soviet Union in Europe, there were little prospects for Russian economic development in Europe. However, Asia might become future world economic center while Russia had a large area of land in Asia. With China's help, Russia might have a prosperous economy in Asia.

The problem was that Russian people were unwilling to go and stay in Russia's Asian part. Russia could issue fix-term visas to Chinese people for them to make contributions to the development in the area, get rich and then return home. In addition, it may absorb those who married Russian girls by stipulating that those who want to marry Russian citizens shall first be nationalized.

That would benefit both Russia and China. China lacks girls as Chinese people prefer having sons while Russia has a surplus of girls.

When the Asian part of Russia has become prosperous, lots of Russian people will move there to look for better opportunities and Russia will no longer have such an urgent need for Chinese immigrants.

If China and Russia help North and South Koreas unify into a prosperous nation, if the ASEAN, the Shanghai Cooperation Organization, the unified Korea and other countries form an economic union, Japan will be forced to join. That will be an economic union larger than the EU.

Putin Tries to Win Advantages from Both Sides in Sino-Japanese Conflict

Putin not only failed to respond to Xi Jinping's request for support in China's dispute with Japan over the disputed islands, but tried to exploit the dispute for Russia's benefit.

In Chinese official media Xinhua's subordinate Asian-Pacific Daily's report on February 7, 2013 about Russian President Putin's meeting with Chinese President Xi Jinping, Putin said that Japanese militarists' war crimes against Asian people should not be forgotten, but that was not reported in Russian English media.

On the contrary, Russia is seeking warmer relations with both China and Japan.

By February 2013, Russian president Putin had met Chinese President Xi Jinping six times since Xi took over the reign in China. When Japanese Prime Minister Shinzo Abe visited Russia to attend the opening ceremony of Winter Olympics in Sochi, Russia, Putin had met Japanese Prime Minister Shinzo Abe five times since Abe became Japanese prime minister.

Abe's decision to attend the opening ceremony came late after all other Western leaders had snubbed Putin by refusing to attend the opening ceremony. Abe did so as he wanted to please Russia, but he certainly displeased the U.S. in doing so; therefore, it was especially pleasing to Putin.

Japan Vital for Putin's Asian Strategy

While forming a Cold War alliance with China to counter the U.S. and preferably drive the U.S. away from Asia, Russia could not but still cherish the fear that according to the trend of China's rapid growth, China will one day be too strong a neighbor.

The common strategy is that as there are three powers Russia, China and Japan in East Asia, whenever one of them is too strong, the other two can form an alliance to strike a balance. Therefore, it is vital for Russia to maintain good relations with both China and Japan so that when one of the two has become too strong, Russia can use the other as a counterweight.

Moreover, at that time, Russia wanted to diversify its energy market. It had concluded an agreement with China on export of oil to China, but the contract on export of natural gas to China could not be concluded after years of negotiation due to disagreement on price. Russia believed that the introduction of a competitor would help resolve the problem. Japan as one of the largest energy importers in the world, was certainly interested in Russian energy resources due to the closeness of their geographical locations. In addition due to the accidents in Japan's nuclear power plant caused by an earthquake, Japan had to switch to natural gas as a clean energy source.

When Japan competed with China for Russian natural gas, Russia would be able to get a better price.

That was why in spite of Xi Jinping's request, Putin could not openly and entirely stand by China's side.

He certainly made China unhappy and his alliance with China not close enough.

184

Some Breakthrough in Sino-Russian Arms Deals due to Putin's Insistence

Russian military worried that providing China with advanced weapons might threaten Russian security while Russian military industry circles worried that China might steal Russian technology by reverse engineering, but Russian President Putin insisted that there should be the arms deals because he wanted a Cold War partnership between Russia and China against the U.S.

Chinese Communist Party's and Russian Putin's autocracies may in essence support each other while the U.S. pivot to Asia constitutes a threat to both China and Russia. Influential U.S. Sixty Minutes Plus's support for Russian youngsters' opposition to Putin's autocracy is a clear example of American people's mindset. As a democracy, U.S. government has to obey the people's will and the media can be dominant.

In spite of the need for closer ties, the negotiation on arms deals made slow progress due to lack of mutual trust between the two parties. In late March 2013, there was media report that informed sources said that China hoped to purchase Russian Su-35 fighter jets because the IRBIS-E passive phased array radar and 117S engine Su-35 uses were the most advanced in the world.

Authoritative Russian military industry top management confirmed China's purchase of Su-35 and regarded it as closer Sino-Russian military cooperation. They believed that the purchase of Su-35 would begin a new wave of Chinese purchase of Russian weapons. The source said, "At present, there are talks on specific technological details and prices; therefore, we have to prepare a formal contract, which maybe will be signed by the end of 2013. Intensive talks are being carried out mainly on what weapon system China needs."

The Su-35 issue is first of all a political decision. Russian source stressed: "All arms deals, especially major arms deals are political decisions." From that issue, we see Putin's urgent desire to form an alliance with China. However, it takes time to build mutual trust.

According to the report, Chinese military had already entered the era of attaching great importance to tactical and strategic air force. Due to lack of technology, Chinese air force needed a 4th-generation engine like 117S. With it, the J-20 fighter jet China was developing would have supersonic cruise capability and be upgraded as a standard 4th-generation fighter.

That seemed to be a sign of renewed contact between China and Russia in the field of radar technological cooperation. The two sides were discussing the possibility of joint development of passive phased array radar. That indicated that

China still had difficulties in developing that kind of radar. If Chinese air force had really obtained Su-35, there would be a further all-round change in the air strategic situation in Asia-pacific region.

Initial Breakthrough in Building Mutual Trust

According to Russian rusnews.cn's report, Alexander Mihyev, vice president of Russian National Defense Export Company, said on June 17, 2013 at Paris Airshow, that Russia did not worry that China would copy Su-35 fighter jet.

In an interview with media, he said, "Russia and China have signed an intellectual property protection agreement. We do not foresee any problem concerning the copying of Su-35."

According to a previous report mentioned in the preceding section, Russia and China is conducting technology discussion concerning supply of Su-35 fighter jets to China. So far, the two countries have an agreement on such export between their two governments.

However due to lack of mutual trust, slow progress had been made in the discussion on weapon deals especially Russian export of military technology to China.

Russia Took the Lead Followed by China to Confront the U.S.

When China received the ball, i.e. Snowden, it was a good ball, but China did not shoot, as it did not dare to confront the U.S.; therefore, it told Hong Kong to pass the ball to Russia on some faint excuse.

Russia, however, liked the chance to show it was able to be the leader of the autocracy camp of the new Cold War it has been forming with China. In spite of potential U.S. fury, it not only welcomed the ball but also dared to shoot. Alas, it scored.

The U.S. was really furious.

Reuters said in its report on August 2, 2013, "'Russia has stabbed us in the back, and each day that Mr. Snowden is allowed to roam free is another twist of the knife,' said Senator Chuck Schumer, a close Obama ally and fellow Democrat who urged Obama to recommend moving out of Russia the summit of G20 leaders planned for St. Petersburg.

"Republican Senators John McCain and Lindsey Graham, already sharp critics of Putin, called Russia's action a disgrace and a deliberate effort to embarrass the United States. They said the United States should retaliate by pushing for completion of all missile-defense programs in Europe and moving for another expansion of NATO to include Russian neighbor Georgia."

However, "bravo," all the autocrats in the world who are anxious to join the autocracy camp of the new Cold War, acclaimed.

China, especially, was glad that Russia was bold enough to be the leader of the camp.

With Russian influence, especially due to their similarity as communist autocracies, China and Vietnam have soon expressed their wish to resolve through cooperation their bitter maritime territorial dispute in the South China Sea and talked warmly about their traditional friendship.

Hong Kong's Singtao Daily reported on August 6, 2013 that in a meeting between Chinese Foreign Minister Wang Yi and Vietnamese Prime Minister Nguyen Tan Dung, Mr. Dung said, "The Vietnamese Party and government will always remember and be grateful to China for China's huge support and assistance in Vietnam's undertakings of national independence, liberation and construction."

What warm statement!

No mention of China's invasion of Vietnam in 1979 or the maritime territorial dispute at all. Obviously, since Russia has taken leadership, the autocracy camp of the new Cold War has been firmly established. Vietnam will certainly be a member of the camp.

Having been openly challenged by Putin, American leaders should not have been surprised by Putin's response to the revolution in Ukraine. Still, they gave people the impression that they were entirely unprepared for the response. No wonder people in the world have lost confidence in U.S. leadership. U.S. President Obama had to visit Asia to reassure U.S. allies there.

16. The New Cold War

Further Trust Needed for Real Cold-War Partnership

The Snowden incident made Putin understand Xi Jinping's intention to let him be the leader of the autocracy camp of the new Cold War. Putin was very happy as the incident made him popular at home and also abroad among the autocratic countries that were members or potential members of the autocratic camp.

That was Putin's first taste of being world leader.

He certainly welcome further test to display his true qualities as Cold War leader to challenge the West and find how strong his allies, especially China, supported him.

Ukraine Test

The Ukraine crisis gave him the right opportunity.

Putin knew well that Xi Jinping was not happy that instead of supporting China in its conflict with Japan, Russia had been actively seeking better relations with Japan as a potential ally and market for its energy export.

Now, China has great interests in Europe as Xi Jinping had got favorable response for his proposal for free-trade relations with EU. Ukraine was even more important for China as it was China's major sources of advanced weapons and weapon technology. Before former Ukraine President Yanukovych was deposed, China had signed contracts worth $8 billion on investment in weapon industry and import of weapons and weapon technologies including advanced aircraft, warship and tank engines, huge transport aircraft, wide steel board for huge warships including aircraft carriers, etc.

If China had refused to take Russia's side for fear of affecting its relations with EU and Ukraine, Putin would have been isolated.

Putin was very happy that as soon as Ukraine crisis emerged, CCP's mouthpiece People's Daily published a commentary on February 27, 2014 that denounced the West's "Cold War mentality" against Russia.

Putin got the message from CCP's mouthpiece that China supported him as he knew very well CCP propaganda machine's view was identical to CCP authority's. That is common knowledge of everyone who has the experience of living under CCP or Soviet Communist Party's rule.

However, Xi Jinping was so skillful to fool people that he made people believe that he remained neutral in spite of his firm support for Putin.

To confirm China's support, in the evening of March 4, Putin talked with Xi Jinping over the phone. The website of Russian President says, "Vladimir Putin and Xi Jinping discussed the exceedingly complex situation unfolding in Ukraine, noting their close positions. They expressed hope that the steps being taken by Russia's leadership will help decrease sociopolitical tension and ensure the security of the Russian-speaking population in Crimea and Ukraine's eastern regions."

Since their positions are "close", certainly China does not think that Russia invasion of Crimea is regrettable. On the contrary, Xi shared Putin's hope that the steps taken by Russia (including invasion of Crimea) would help decrease tension and especially "ensure the security of the Russian-speaking population" in eastern Ukraine.

Chinese official media CCTV's primetime report on the telephone conversation between the two presidents on March 4 refrains from openly mention the "close positions" but gives hints on that. It says, "Xi Jinping expounded China's principled stand and pointed out there were inevitability in the fortuitous development of the situation in Ukraine to what it is today."

What did Xi mean by "inevitability"? He hinted that Western interference made the fortuitous events inevitably develop to what it was at that time.

Xi hinted that he shared Putin's position by saying that he "believes the Russian side will be able to coordinate with various parties, promote a political settlement of the issues and safeguard regional and world peace and stability."

Chinese Tricks in Appearing Neutral While Supporting Russia

In order to hide China's support for Putin and appear neutral, Chinese officials played the tricks to give replies to reporters difficult to understand.

To understand that, first we have to make accurate translation of their words. Then we can analyze the meaning between lines. As I have been a legal translator for the largest transnational law firms for decades, I have the expertise to do so.

The following is the accurate translation of related passages in the records of Foreign Ministry's routine press conference on March 4 published on the Ministry's website:

Question: China has expressed its always adhered principle of non-interference with internal affairs and at the same time also takes into consideration of the course of developments and the external relations in history as well as the complexity of the reality. What "the course of developments and the external relations in history" means specifically? Does China deny that Russia's action in Crimea is interference with Ukraine's internal affairs?

Answer: Regarding Ukraine issue, China has already made clear its position. As for the course of development and the external relations in history you mentioned, please review and refer to the relevant history of Ukraine and that area. I believe when you know the relevant history, you will understand what we meant when making clear our position.

Regarding the second question, I would like to ask you to understand China's position in a complete and systematic manner. We adhere to the principle of non-interference with internal affairs and respect international law and the generally accepted norms of internal relationship and at the same time take into consideration of the course of development and external relations in history and the complexity of the reality. You can make an analysis of the reasons why the Ukraine issue has developed to what it is today from various parties' activities and conducts during the past few months.

The meaning between lines:

Taking into consideration of the history and reality, Russia's activities and conducts in Crimea are justified; therefore, though China opposes interference with internal affairs, China does not oppose Russia's actions and conducts in Crimea.

However, China opposed the West's interference. China had double standards for West's and Russia's actions and conducts due to its consideration of the course of development and external relations in history and the complexity of the reality!

The meaning between lines will enable people to see clearer the situation of the new Cold War of Russia plus China v. the United States.

Regarding the question again whether China regards Russia's interference in Crimea as regrettable. China always advocates none-interference with other country's internal affairs, but Crimea was an exception.

Chinese official played the trick again at the routine press conference of Chinese Ministry of Foreign Affairs on March 7. The spokesman Qin Gang gave comments on U.S. and EU's view that the referendum on Crimea's merger with Russia is a breach of international law.

The following is the precise translation from Chinese of the question and answer:

Question: It is reported that the Parliament of the Autonomous Republic of Crimea has announced to hold a referendum within 10 days on whether to merge with Russia. The United States and EU regard it as a breach of international law. Does China have any comment on that?

Answer: We call on all the parties concerned <u>in Ukraine</u> to <u>peacefully</u> resolve the relevant issue within a legal and orderly framework through dialogue and negotiations, earnestly safeguard the legitimate rights and interests of the people of various ethnic groups in Ukraine, recover normal social order as soon as possible and maintain peace and stability in this region (My underline for understanding the meaning between lines).

The meaning between lines:

1. China wants a <u>peaceful</u> solution among various parties <u>in Ukraine</u> as it is the country's internal affairs; therefore, there is no question of whether the referendum breaches international law or not!

2. It implied that both the U.S. and EU are interfering with other country's internal affairs. China always advocates non-interference of other country's internal affairs.

3. What about Russia? Are Russian Troops not in Ukraine to interfere with Ukraine's internal affairs?

Russia denies the troops in Ukraine are its troops and China believes Russia's lie to support Russia. China is certainly not so naïve as not to know the truth.

4. When it has been decided by the referendum that Crimea shall merge into Russia, the issue shall be resolved peacefully, i.e. China holds that the Ukrainian troops shall not attack Crimea if Crimea has been annexed by Russia.

The translation of the question and answer about sanctions:

Question: According to reports, the United States and EU are discussing the imposition of sanctions against Russia. Does China supports sanctions against Russia?

Answer: China has consistently opposed being too ready to impose sanctions or using sanctions as a threat in international relations. In the present situation, we hope that all the parties concerned can avoid taking any action to further intensify the tensions and make joint efforts to find a way for the political solution of the crisis. This is the fundamental way out.

The meaning between lines:

1. China opposes the sanctions. In order not to offend U.S. and EU, it explains that it is its consistent stance. It has nothing to do with the issue or the parties concerned; therefore, it shall not offend the U.S. or EU.

It pleases Russia as China opposes sanctions and certainly will not take part in the sanctions.

2. China hopes that all the parties can avoid intensifying the tensions. It implies that sanctions will intensify tensions and shall, therefore, be avoided.

3. China advocates political solution, hinting that neither U.S. and EU nor Russia shall seek a military solution. It implies that Russia now is seeking a political solution as it has denied that it has sent troops into Crimea and China believes that.

Then what if Russia sends troops into other parts of eastern Ukraine and the parliaments there hold referendums to decide merger with Russia? China advocates a political instead of military solution, i.e. the U.S., the EU and even Ukraine shall not send troops to resolve the issue.

Xi Jinping's Skill in Subduing the Enemy with Diplomacy

Allying with Russia to counter U.S. pivot to Asia is a strategy initiated by Hu Jintao and inherited by Xi Jinping. With that strategy, Hu succeeded in breaking U.S. encirclement of China, but China was still unable to exploit the rich resources in the South China Sea due to U.S. support for Philippines and Vietnam's contending claims.

China's land area is but 9.6 million square km, but the South China Sea claimed by China as being inherited from its ancestors has an area of 3.5 million square km and is rich of energy and fish resources. Grabbing back the land territories China lost in the past is regarded as aggression and may lead to wars with powerful neighbors, but maintaining the sea areas claimed by Chinese ancestors are not so heinous. The disputes over the sea areas are merely regarded as border disputes. All the countries that are not parties to the disputes such as Russia, EU members and Latin American and African countries just ignore the issue.

The United States certainly is afraid that with such a vast sea area added to China's vast land area, China will be even greater a rival to it in the competition for the number one status in the world; therefore, despite its traditional attitude of taking no side in other countries' border disputes, it want to interfere. It has been weakened by the two wars in Iraq and Afghanistan and the financial crisis in 2008, but remains habitually arrogant and wants to interfere with other countries affairs. It has been using its pivot to Asia to interfere with China's disputes in the South China Sea.

Therefore, China has first of all to crush U.S. arrogance without damaging its relations with the U.S. Russian President Vladimir Putin is precisely the person capable of and interesting in challenging the U.S. while the Ukraine issue precisely provided Putin with such an opportunity.

Sino-Russian Alliance

An alliance must first of all be mutual beneficial. Certainly, it is unavoidable that the two parties to an alliance may have conflict of interest, but the conflict shall be insignificant compared with the benefit.

That is the case with Sino-Russian alliance. It is mutual beneficial because:

1. Russia and China need each other as its ally to counter the U.S. especially Russia at that time had lots of troubles in Europe. Development of the rich resources in its vast Asian area seemed the only way out. U.S. pivot to Asia seemed directing at China, but for Russia, it also directed at Russia. Both Russia and China were much weaker than the U.S., but their combined strength is enough to counter the U.S.

2. Russia relies on its energy export, but as nearly all its west neighbors want to join EU, the route to Europe may well be blocked. It had better export energy to China and Japan, the two biggest oil and gas importers in the world. There is no one to block Russia's export to the east. For China, its import of oil from the Middle East and Latin America may be blocked by U.S. navy that dominates the oceans. Getting oil and gas from Russia is much safer.

3. China wants Russian advanced weapons due to its lack of technology while Russia needs the proceeds of its weapon sales to arm itself in order to recover its position as a rival to the U.S.

4. China may contribute funds and labor for Russia's development of Siberia and get resources in return. This will be a bright mutually beneficial future for Sino-Russian relations.

However, the rise of China may become a threat to Russia; therefore, Russia has been making great efforts to improve relations with Japan. The three Asian powers China, Japan and Russia will constitute a balance of force in Asia. When one of them is too strong the weaker two may form an alliance as a counterweight.

Such alliance results from necessity and interests. It has nothing to do with love or affection.

China's close relation with the U.S. before U.S. pivot to Asia, however, is a relation out of love. China had lots of interests in Libya, but did not hinder Western military intervention there. China had to withdraw 50,000 citizens from

Libya and suffer great economic losses.

Still the U.S. took no account of that and began pivot to Asia to contain China. I regard it as China's love for the U.S. at that time, but it was an unrequited love.

What about hatred? China may hate Russia as it lost 2 million square km of land to Russia a century ago. However, the example set by France and Germany proves that such historical issue shall better be forgotten forever. Otherwise, there will be endless military confrontations and wars between the two neighbors.

In fact, as China and Russia have signed border agreement to permanently resolve their border issue, media may invent some sensational story about Sino-Russian border conflict, but there is no actual dispute whatsoever.

However, the Sino-Russian ally is not a very close one. China does not want to offend the West. It wants to have a free trade agreement with EU for economic benefit. Its export to the U.S. is providing lots of jobs for China's huge population.

Russia, on the other hand, has not shown any clear or strong support for China in China's conflicts with Japan. On the contrary, Russia exploits the conflicts to improve relations with Japan.

Now it is China's turn to win advantages from both sides. The tension in Ukraine will greatly reduce the pressure of U.S. pivot to Asia on China; therefore, being wise, Chinese leaders' strategy will of course be supporting Putin in private to keep the tension there. However, in order not to offend the West and to prevent the West from giving up due to strong Sino-Russian ally, China has kept its stance quite vague in public but quite clear to Russia in private. If one does not know how to read between lines Xi Jinping's words in his telephone conversation with Putin and Chinese foreign ministry spokesman's words on Ukraine, one will remain puzzled what China's attitude really is.

How Xi Jinping Pleased U.S. and the West while Supporting Russia

China is obviously on Russia's side, but it does not offend the U.S. and EU by that. Xi Jinping even pleased U.S. President Obama and German Chancellor Merkel. We can clearly see that in U.S. and Germany's responses to what Xi Jinping said to Obama and Merkel in their telephone conversations.

U.S. Office of the Press Secretary said in its "Readout of the President's Call with President Xi of China" that in the telephone conversations, "they affirmed their shared interest in reducing tensions and identifying a peaceful resolution to the dispute between Russia and Ukraine. The two leaders agreed on the importance of upholding principles of sovereignty and territorial integrity, both in the context of Ukraine and also for the broader functioning of the international system."

Wield! Xi Jinping was able to maintain a position close to Russia's while sharing interest with the U.S. Were Russia and the U.S. not confronting each other over Ukraine at that time?

What did Xi said to give the U.S. such an impression?

Let us analyze what Xi Jinping said to Obama on Ukraine via telephone according to CCTV primetime report.

The following is an accurate translation of the relevant part of the report:

"Obama briefed Xi Jinping America's views on current Ukraine situation. Xi Jinping stressed that China adopted an objective and just attitude towards the Ukraine issue. What is urgent for all the parties to do now is to keep calm and restraint and avoid further escalation of tensions in the situation. They shall persist in resolving the crisis by political and diplomatic means. It is hoped that the various parties concerned satisfactorily deal with relevant differences through communications and coordination and make efforts for a political resolution of the Ukraine issue. China adopts an open attitude to all suggestions and schemes that facilitate easing the tension in Ukraine and is willing to continue to maintain communications with the U.S. and all other parties concerned."

The U.S. was afraid that due to China's interests in investment and weapon technology in the contracts it signed with former Ukrainian President Yanukovych, China might oppose Ukraine's new government. However, the U.S. could rest at ease that at that time China upheld an objective attitude in disregard of its interests.

In addition, due to China's alliance with Russia, the U.S. is afraid that China may be on Russia's side especially Russia said Chinese and Russian presidents' held close positions. However, since Xi said that China adopted a just attitude, China is certainly not on Russia's side.

However, if we read between lines, we find that when Xi heard Obama's view, Xi stressed China's just and objective attitude. He hinted that the U.S. failed to adopt a just and objective attitude.

It is especially so as in other occasions, China said that it adopted an objective and just attitude, but now it stressed its just and objective attitude. The change in order to place just before objective stressed that U.S. attitude is not just.

Russia will certainly be happy when Chinese diplomats explain the meaning between lines to it.

Xi Jinping has also made Germany happy. Reuter says in its report "China's Xi urges political solution to Ukraine crisis", "'The chancellor explained the situation in Ukraine and efforts to come to a political solution of the conflict,' German government spokesman Steffen Seibert said in a written statement.

"'The Chinese president was also in favor of finding such a solution through dialogue,' the statement said, adding that Xi said the solution needed to be on the basis of international law."

The following is an accurate translation of Xi's words in the telephone conversation according to CCTV's report:

> Xi Jinping pointed out: Current Ukraine situation is very complicated and highly sensitive. It is necessary to examine, consider and weigh it in its entirety in dealing with it. China is also following the situation closely. We call on all parties to keep restraint, achieve a political resolution of relevant differences within a legal and orderly framework through dialogue and negotiations and prevent the situation from escalating. China supports international community's constructive efforts and actions of mediation in easing up the tension. He hopes that the German side will keep on communicating with other parties concerned and give further play to its constructive role. The Chinese side is willing to maintain contact with the German side.

What Xi said certainly pleased Merkel, especially on resolution within legal and orderly framework. For Merkel, it meant that Russia had violated international law; therefore, should withdraw its troops according to international law and Crimea's referendum on merger with Russia violates international law.

However, it did not offend Russia at all as Russia said that it had not sent its troops into Crimea and in Xi and Putin's telephone conversation, Xi told Putin, "he believes Russia will help resolve the issue through political means and maintain regional peace and stability". Since Xi believed Russia would help resolve the issue through political means, he certainly believed Russia had not sent troops into Crimea.

Moreover, Xi's requirement for examining, considering and weighing the situation in its entirety meant besides respecting Ukraine's territorial integrity, Crimean people's right of self-determination should certainly also be respected. That was precisely what Putin wanted.

Sending troops to take Crimea away from Ukraine violated international law, but allowing Crimea to merge with Russia according to Crimean people's self-determination did not violate international law.

It was very difficult for Xi to maintain his position close to Russia's while avoiding to offend the U.S. and EU, but he had managed to achieve that.

Such skillful handling of the very complex and sensitive issue reflected Xi Jinping's marvelous wisdom.

China Even Pleased Ukraine in Spite of Being Close to Russia

Chinese Foreign Minister Wang Yi said that development of the situation in Ukraine to what it is today is regrettable. Why? China regrets that China's interests may be gravely harmed as China has signed lots of agreements involving $8 billion investment when former Ukrainian President Viktor Yanukovych visited China last December especially for import and joint development of advanced weapons and technology. With those agreements, China will get the technology to build the largest transport aircraft in the world, advanced engines for its light stealth fighter jet, the technology for production and welding of wide thick steel plates for the new aircraft carriers China is building, etc., which are vital for China's military modernization. In addition, China has already had investment in about 10% of Ukraine's farmland. All those may be seriously affected.

However, there is no need to regret later. In his recent interview with Global Times reporter Li Qian, Ukrainian Ambassador to China Oleg Dyomin said, "The cooperation between Ukraine and China is based on the principle of mutual benefit. We never had disputes in economic or other forms of cooperation. The new government of Ukraine, upon taking power, immediately notified China it would continue all contracts between the two sides."

Why? Dyomin said, "Here we want to emphasize that Ukraine values China's consistent stance on supporting Ukrainian territorial integrity and independence."

While giving Russia the impression China's position is close to Russia's, China had given Ukraine the impression that it was on Ukraine's side and supported Ukrainian territorial integrity. That was really wonderful.

Ukraine "hope China, one of the world's leaders and a partner of Russia, will assist with talks between Ukraine and Russia as an intermediary to tackle issues through negotiations," Dyomin added.

Now, China's close position to Russian is an advantage in maintaining good relations with Ukraine as China can play the role of intermediary.

What about Russia? Has it been offended by China's attitude towards Ukraine?

Not at all. According to Global Times reporter Li Qian's exclusive interview with Russian Ambassador Andrey Denisov, Denisov believed that recently the cooperation between China and Russia has reached "new highs" "in all fields."

He said, "Leaders and foreign ministers of our two countries have discussed the current crisis in Ukraine. We are very grateful to China for its balanced and principled position on this issue."

With China by its side, Russia has grown bolder in its attempt to recover its satellite countries. It annexed Ukraine's Crimea in early 2014 and no one knows whether it will annex Luhansk and Donetsk areas that have held a referendum to split from Ukraine. I believe the two areas will at least become Russia's satellite states. With Chinese support, Russia will gradually take back some parts of the old Soviet Union and the U.S. can do nothing to stop Russia.

China Helps Russia by Opposing Western Sanctions

At the critical moment when the West is deciding to impose sanctions on Russia, China came out to help Russia. It gave the West the advice to be patient and continue to solve the issue through dialogue.

What it really wants is to join Russia in opposing the U.S. due to U.S. pivot to Asia while maintaining good relations with Europe.

To cover up China's support for Russia, on March 12, 2014, China's top envoy to Germany used the excuse that punishing Russia with sanctions for its intervention in Ukraine could lead to a dangerous chain reaction that would be difficult to control.

"We don't see any point in sanctions," Ambassador Shi Mingde said in an interview with Reuters. "Sanctions could lead to retaliatory action, and that would trigger a spiral with unforeseeable consequences. We don't want this."

Shi urged patience, saying the door for talks should remain open even after a referendum in which Ukraine's southern region of Crimea could vote to secede and join Russia. German Chancellor Merkel and other western leaders had denounced the referendum as illegal and demanded that it be canceled.

"We still see a chance to avoid an escalation. The door to talks is still open. We should use this possibility, also after the referendum," Shi said.

Also for covering up China's support for Russia, President Xi Jinping said on March 28, 2014 at a joint press conference with German Chancellor Angela Merkel that China would not take sides with the West or Russia over Ukraine, disappointing any hopes Beijing might add its weight to international pressure on Moscow for annexing Crimea.

Xi said, "China does not have any private interests in the Ukraine question. All parties involved should work for a political and diplomatic solution of the conflict."

As a matter of fact, China had the interest in joining Russia in opposing the U.S., but anxious to have good relations with EU. Xi certainly was unwilling to disclose that.

On April 28, 2014, China restates opposition to sanctions on Russia over Ukraine after leaders of the Group of Seven (G7) major economies agreed to swiftly impose further punitive measures.

Chinese Foreign Ministry spokesman Qin Gang repeated China's excuse that sanctions were not conducive to resolution of the issue, and may worsen tensions.

China Proved Itself Russia's Trustworthy Cold War Ally

China proved itself Russia's trustworthy Cold War ally by its skillful opposition to West sanctions against Russia and refraining from denouncing Russia for what it did in Ukraine.

Russian President Putin was so pleased that before his Shanghai visit on May 20-21, 2014, he said in an interview to major Chinese media including Chinese Central Television, Xinhua news agency, China News Service, The People's Daily and China Radio International, "Now Russia-China cooperation is advancing to a new stage of comprehensive partnership and strategic interaction. It would not be wrong to say that it has reached the highest level in all its centuries-long history."

Putin meant: Russia and China are closer allies than they were in the early 1950s when Russia was a part of the Soviet Union. However, there was a treaty of alliance between China and Soviet Union at that time, but there is none now. But perhaps treaty is not so important for the allies. In the 1960s, China and the Soviet Union became bitter enemies in spite of the treaty.

In order to win Russia's support in China's dispute with Japan, Xi Jinping said in a speech he gave when he visited Russia in March 2013, that China and Russia should jointly preserve the achievements of the victory of World War II and maintain the international order after the War.

Xi meant that as Russia got the four islands in dispute with Japan due to World War II, China supported Russia in the dispute. On the other hand, as Japan returned Taiwan to China due to its defeat in World War II, it shall give back to

China the Diaoyus (known as Senkakus in Japan) as the Diaoyus are a part of Taiwan while Taiwan is a part of China. However, Russia failed to respond as Putin was trying to improve Russian relations with Japan.

However, due to closer alliance, this time Putin gave favorable response.

To reporter's question: In 2015, our countries will celebrate the 70th anniversary of Victory over fascism. What is the impact of joint Russian-Chinese efforts to oppose the attempts aimed at challenging the results of World War II?

Putin said in his reply. "Next year we will hold a range of joint events to mark the 70th anniversary of Victory both in the bilateral and the SCO format. During these events, youth will be in the focus of our work.

"We will certainly continue to oppose attempts to falsify history, heroize fascists and their accomplices, or blacken the memory and reputation of heroic liberators."

What he said has been written into the first part of his joint statement with Chinese President Xi Jinping on May 20, 2014.

Putin Declared Cold War against the U.S.

Having established Cold War alliance with China, with the Shanghai Cooperation Organization (SCO) and Conference on Interaction and Confidence-Building Measures in Asia (CICA) behind him, Putin declared Cold War against the United States in his speech on May 23, 2014.

His typical denial of U.S. leadership is shown by what he said about U.S. President Obama: "Who Made Him a Judge?"

His statements that the U.S.-led world order "has failed" and that, "The world is really changing rapidly. We see colossal geopolitical, technological and structural shifts. The unipolar model of the world order has failed," meant that the U.S. could no longer dominate the world and there were Russia and the Cold War camp led by Russia against the U.S.

Having Nullified U.S. Pivot to Asia by Diplomacy, China Takes Bold Offensives in the South China Sea

One of China offensives was moving its huge oil rig to disputed waters with Vietnam. Previously, China gave the impression that it would not exploit the rich oil and gas resources in the South China Sea until its disputes with other claimants had been resolved. Now China's offensive indicates that it will exploit the resources in spite of other claimants' opposition.

Another serious move is its construction of a military base at Johnson South Reef. According to the information provided by Philippine government, China has already reclaimed 0.09 square km of land there.

Earlier, there was news on January 11, 2014 that Chinese navy has drawn detailed combat plan to seize Zhongye Island back from the Philippines. As the remote area of South China Sea is too far away for Chinese fighter jets to cover, China has to set up an air base somewhere in the middle of the sea. Zhongye Island seems a good choice due to its location. In addition, building an air and naval base there costs much less than building an aircraft carrier. U.S. carrier the *Ford* costs $11.2 billion to build but only has a deck area of 0.026 square kilometer while the Island base will cost much less and has a land area much bigger (0.33 square km). And it is unsinkable.

However, China had to consider the harm the attack might do to its relations with the U.S. and ASEAN; therefore Chinese central authority did not approve Chinese navy's plan.

I shall praise Chinese leaders' wisdom that they pay attention not only to the strategic military significance of their islands in the South China Sea but also their economic benefits. As a result there was news later about China's plans on building a military base on Fiery Cross Reef and a fishing and fish-farming base on Mischief Reef so that China can recover the construction costs from the income of the fishing base.

When I learnt China's ownership of the remote uninhabitable islands and reefs in South China Sea in my early childhood, I wondered: what are the benefits in having those uninhabitable islands and reefs. I thought they were burdens as China had to keep a navy to defend them. Later, it turned out that there were rich oil and gas resources in the sea around them. How lucky China is!

Now, when China's rivers are polluted and sea areas near China, not much better. It is indeed a wonderful idea to use the islands and reefs as unpolluted fish farms to satisfy Chinese people's huge demand for fish.

China is conducting reclamation at Johnson South, Cuarteron, Hughes, Mckennan, Gaven and Eldad Reefs.

It seems that Johnson South Reef is better for a military base as it is 2.4 km wide and 3.8 km long. By reclamation of its lagoon, China can easily get an area much bigger than Zhongye Island. The Reef is located also in the middle of the South China Sea area China claims as its own, and a little further south than Zhongye Island, better for control of the remote southern area.

However, by May 2014, the artificial island on Cuarteron Reef turns out to be the biggest among those built on the six reefs. It may also be used as a military base. The reclamation on the other four reefs, I guess, is for construction of bases for fishery, fish farming and tourism.

That will bring about a tremendous change in the situation there. The military bases, if completed, will have a well-equipped airport and a naval and a coast guard base. The Philippines and even Australia will be within the range of China's medium-range and intermediate missiles and bombers.

When the fish farming has proved profitable, China will develop all the large number of uninhabited reefs there.

Precious Legacy

What is the precious legacy left by Chinese people's predecessors? The vast land, the culture, philosophy, knowledge, technology, experience in their long history, etc.

What people often neglect is the vast sea area within the nine-dash line.

What is the use of the waters? No crops can grow there. Fishery, fish farming and alga farming may be more profitable than agriculture on the small proportion of farmland on China's land area, which is mostly deserts, mountains and very high plateaus. Only less than 10%, i.e. 0.96 of the 9.6 million square km of China's land area can be used for farming.

The sea within the nine-dash line has more than 3 million square km that can all be used for fishery and fish and alga farming.

In addition, we all know that there are rich oil and gas resources there.

However, we still do not know what rich mineral resources there are at the bottom of the sea.

Chinese people certainly have to keep the rich legacy left by their ancestors.

Shall China Fight Six Wars for "Reunification"?

> Subduing the enemy by stratagem is the best; by diplomacy, second best; by battles in the field, the third alternative; by attacking enemy cities, the last resort.

> --*The Art of War* by Sun Tzu

Fighting and winning each and every battle is not the best of
the best; subduing the enemy without fighting is the best of the
best.

–ditto

Amid surge of nationalism due to maritime territorial disputes in the East and
South China Seas, a naïve girl Li Qiuye wrote an article entitled "Six Wars China
Is Sure to Fight In the Next 50 Years". Unexpectedly, the foolish article was quite
popular and lots of Chinese media accept Li's views.

The six wars are: The war to take Taiwan for reunification; with Russia for
recovery of 2 million square km ceded to Russia by a treaty, the war with
Mongolia to annex it as it was a part of China before 1924 and gained
independence due to Soviet influence; the war with India for disputed land areas
some 100,000 square km, the war with Japan for disputed uninhabited small
Diaoyu (known as Senkaku in Japan) Islands; and the war for disputed islands in
the South China Sea that may involve Vietnam and the Philippines.

Li regards the six wars as wars for reunification, but only the war with Taiwan
was really for reunification while the wars with Russia and Mongolia are for
recovery of lost territories and those with Japan, India, Vietnam, etc. are to resolve
territorial disputes.

France and Germany fought for nearly a century for the disputed
Alsace-Lorraine area, which was even one of the causes of world wars. Neither
France nor Germany is so stupid as to fight on. They finally decided to set up the
EU to put an end to the dispute forever.

Is China wise to fight wars to turn most of its neighbors into its enemies?

What will China get even if it wins all the wars?

First, the most important issue–the reunification with Taiwan. Reunification by
force will bring China no benefit as Taiwan has little resources and no cheap labor.
The damage done by the war will make China not only lose a substantial market
but incur the heavy burden of feeding 20 million Taiwan people and maintaining
their high living standards. Moreover, Taiwan is an island full of hills favorable for
guerilla wars. The financial burden and the guerilla wars will make the Chinese
regime unpopular and even collapse.

Li does not understand that the driving force is interest instead of the lofty
slogan of reunification.

Russia, Mongolia, India and Japan will make every effort to recover the territories taken by China. There will be no end of wars with them. China will get nothing from those recovered barren territories and islands, but will incur lots of expense in defending the recovered territories.

Obviously the third alternative battles in the field and the last resort of attacking cities will not bring any benefit to China.

Taking Back Islands and Waters by Diplomacy

Of the six wars, four will be fought against or involve very strong countries: Russia, India, Japan and the United States who has obligation to protect Taiwan. If China fought the four countries within the said five decades, China will have troubles for centuries.

Taking back the islands and waters in the South China Sea is quite different. The countries that contend with China for the islands and waters are all small and militarily weak. Even if China fight wars with them to take back the islands and waters, China will not have much trouble after the wars.

The country that may interfere is the United States, which has adopted its pivot to Asia to contain China, but China has so far successfully countered it with the establishment of the above-mentioned Cold War alliance with Russia.

Some media outside China are fond of listing Taiwan, Vietnam, the Philippines, Malaysia and Brunei as contenders for the islands and waters. They treat Taiwan as a country separate from China as they hope so to make China less strong; therefore, they just ignore the fact that both China and Taiwan hold that there is only one China. Taiwan regards itself as China and the China on Chinese mainland as a part of it while China regards itself as China and Taiwan as a part of it. If those contenders other than Taiwan are willing to return to Taiwan what they have taken from China, China will certainly welcome such moves. Then the maritime territorial disputes with those contenders will become an issue of the reunification of China.

Certainly, however, those contenders will never take such moves.

It is easy for Chinese navy to defeat all the navies of Vietnam, Malaysia and the Philippines at the same time to recover those islands and waters, but that will be the third alternative.

"Subduing the enemy without fighting is the best of best," said Sun Tzu in his *The Art of War*.

I believe that China shall certainly choose diplomatic instead of military solution of the problems. It shall first reach an agreement for sharing resources with one or two of the four contenders. It seems that Malaysia and Brunei may be willing to accept such a solution. With the satisfactory examples set by the two contenders, Vietnam under Russian influence may follow suit. Then there will only be the Philippines left. Simple naval blockade of the islands occupied by the Philippines will do. There is no need to fight.

The South China Sea involves too great interests, China certainly shall not cede even one inch of it. However, as alliance with Russia may prevent U.S. interference, China can resolve the issues at will. It needs not wait till it has integrated space and air capabilities to surpass the U.S.

The U.S. Underestimated Russia-China Alliance

U.S. Secretary of State John Kerry said at Bandar Seri Begawan, Brunei on July 1, 2013, U.S. "actions are not intended to contain or to counterbalance any one country". It indicated his ignorance of the Russia-China alliance being formed at that time; therefore, he thought that the U.S. was dealing with only one country. At that time, the U.S. had to deal with two countries, China and Russia, and perhaps an autocracy Cold War camp to be jointly established by Russia and China that will contain Cuba, Vietnam, North Korea and other autocracies.

By providing China with advanced weapons and technology, Russia will get lots of funds from China to build up its military to recover Soviet Union's position as a rival to the U.S.

In addition, Russia and China have been carrying out their annual joint military drills on an increasingly larger scale. Russia has taken offensive in Europe and got a part of Ukraine while China is taking offensives in the South Chine Sea. The U.S. is now helpless in the face of Russia-China alliance.

Certainly, the U.S. will remain the number one in the world. Only there may be another number one Russia. That will be the recovery of the Cold War situation.

China will be able to continue its peaceful rise without U.S. interference. It may become world number one economically, but shall not contend with the U.S. or Russia for world number one.

When it has obtained integrated space and air capabilities superior to U.S. Air-Sea Battle, it will no longer be bullied by any superpower including the United States.

Putin's Plan to Add a New Ally to Russia-Chinese Alliance

Having satisfied with Russia-Chinese de facto alliance regarded by Putin as closer than ever, even than the Soviet-Chinese alliance in the 1950s, Putin wants to expand the alliance to overwhelm the U.S. In order to realize his ambition for the restoration of Russia's status of the former Soviet Union as a superpower rival to the U.S., Putin plans to add India to the alliance. As India is now a rising emerging economy, an alliance of Russia, China and India will be even more powerful, enough even to drive the U.S. away from Asia.

Chinese history provides much experience of one strongest state forming separate alliance with weaker ones to conquer them one by one and the weaker ones forming alliance to resist the strongest one.

The U.S. perhaps lacks such experience due to its short history so that it fails to use China as an ally to deal with Russia when China has "unrequited love" for it. On the contrary, it turned China away to Russia's side by its pivot to Asia. As a result, Russia and China gradually formed an alliance that has given rise to the West's trouble in Ukraine and China's offensives in the South China Sea.

Media outside China misunderstood China and regarded China as more assertive before China began to take offensives. In fact, China had remained defensive before. The Scarborough standoff took place because Philippine navy began to capture Chinese fishing boats and forbid Chinese fishermen's fishing there. It was not a case of China being more assertive as described by the media outside China.

Due to U.S. pivot to Asia, the Philippines began to take offensive thinking it could rely on U.S. support. Unfortunate for the Philippines, the U.S. failed to support it. As a result, China took complete control of the Scarborough Shoal and forbid fishing by Philippine fishermen there.

Even with such success, China had refrained from adopting its navy's plan to take Zhongye Island back from the Philippines for fear of harming its relations with the U.S. and ASEAN.

Now, as China has Russia as a reliable ally and as the West is in trouble in Ukraine, China is taking bold offensives at the South China Sea. It sent a huge oil rig to drill at the disputed sea area with Vietnam, which signaled the beginning of its oil and gas exploitation in disputed waters. It began intensive construction of artificial islands on six reefs in the disputed waters in the South China Sea. If such islands are used as naval and air bases in the middle of the South China Sea, China will be able to control the disputed waters and counter U.S. presence at Philippine military bases.

It has now dawned on the U.S. though perhaps too late that other countries are forming an alliance against it, the U.S. has to make some counter moves. The key now is India.

There are problems for the three-country alliance:

First, who shall be the head of the alliance? I have mentioned above that Chinese President Xi Jinping passed Snowden to Putin to give Putin a clear signal that Xi wanted Putin to be the leader in confronting the U.S.

As India is not strong enough to be the leader, Putin can be certain that he will be the leader of the alliance.

Second, the much trickier problem is the border disputes between China and India.

This time, the U.S. is wise to draw Modi, the newly elected Indian prime minister, to its side. The U.S. has invited him to visit the U.S. in September to drive a wedge between India and China.

Putin is certainly shrewd enough to be aware of that. Due to prolonged good relations between Russia and India, Putin has certainly been doing his best to urge India and China to conciliate.

Xi Jinping is shrewd too. Soon after Modi was elected, he dispatched his foreign minister Wang Yi as his special envoy to visit India on June 8 and 9, 2014.

According to Wang Yi, his visit is an unqualified success.

In its report, Bloomberg quotes Wang Yi as saying to reporters at the end of his two-day visit including a meeting with Indian Prime Minister Modi, "Through years of negotiation, we have come to an agreement on the basics of a boundary agreement, and we are prepared to reach a final settlement."

What is even more worrisome for the U.S. is Wang's description of "China-India cooperation as a massive buried treasure waiting to be discovered," Wang said. "The potential is massive."

India is not less enthusiastic. Its Foreign Ministry says in its website, India "PM emphasized the potential for greater cooperation between India and China for a strong and prosperous Asia, working for mutually beneficial trade and investment as economic partners, joining hands in various areas like counter-terrorism as neighbors, and promoting vigorous cultural exchanges as inheritors of ancient civilizations having extensive historical and spiritual contacts."

Reuters said in its report on June 10, 2014, "Indian Prime Minister Narendra

Modi urged greater cooperation with China on Monday (June 9, 2014) and said he planned to visit Beijing soon, underlining his administration's promise to make a new beginning with the country's giant neighbor."

U.S. Vice President Biden attaches great importance to relations between country leaders. He has contracted friendship with Xi Jinping since Xi visited the U.S. as China's vice president. The friendship facilitates his success in his later visit to deal with China's establishment of the East China Sea Air Defense Identification Zone (ADIZ). The visit seemed a failure, but was in fact a success perhaps due to the friendship. At that time, China was provoking Japan to fight a war with it by the establishment of the ADIZ, but Biden prevented the war by obtaining China's promise not to fire the first shot. As the U.S. is able to persuade Japan not to fire the first shot, there has so far been no war in spite of the tension.

U.S. relations with Modi are precisely the opposite. While China has cultivated satisfactory friendship with Modi by treating him as state head during Modi's four visits to China as a weak opposition leader, U.S. leaders did not have the discerning eyes to tell greatness from mediocrity. The U.S. had snubbed Modi for a long time until he was elected Indian prime minister.

Compared with the U.S., China's border dispute with India is a great problem, but the potential benefits of economic cooperation are a counterweight to offset the problem. There are no such benefits in U.S.-Indian relations. Therefore, China is at a much better position than the U.S. due to its leaders' personal relations with Modi.

Biden has the charm to contract friendship with other countries' leaders that Obama does not seem to have. However, Biden does not seem hopeful to be the next president.

Anyway, the U.S. has to make great diplomatic efforts as Japanese Prime Minister Shinzo Abe has been trying hard to improve relations with Russia. Abe has met Putin five times since he was elected. Only one time less than Xi Jinping since Xi became Party General Secretary. Putin wants to sell Japan while Japan wants to buy Russian oil and gas. There are great interests involved.

On the other hand, if Xi Jinping's reform gives rise to Chinese people's huge demand for foreign goods, Japan will be America's fearful competitor for the Chinese market. Under such circumstances, Japan may very likely be drawn to Russia's side. If that happens, what ally will the U.S. have in Asia? Only the ungrateful Philippines?

It seems the U.S. has to learn from Sun Tzu's teaching:

Subduing the enemy by stratagem is the best;
by diplomacy, next best;
by fighting in the field, third alternative;
by attacking cities, last resort.

--The Art of War by Sun Tzu

The U.S. seems to only know the third alternative, but without stratagem and diplomacy, it cannot win even by the third alternative no matter how strong it is militarily. The U.S. suffered thousands of casualties and spent almost $1 trillion in its war in Iraq. Now, its enemy in Iraq has come back with a vengeance and the government it has helped to set up there seems on the verge of collapse.

Will Iraq be another Vietnam?

My advices to the U.S.: Switch to developing space and air capability so as to remain on top militarily. The world needs the U.S. as a pillar of peace and democracy.

ABOUT THE AUTHOR

This is the author's second book after recent publication of the expanded 2nd edition of his first book *Tiananmen's Tremendous Achievements* to commemorate the 25th anniversary of Tiananmen Protests.

The author Chan Kai Yee was born in a successful doctor's family in 1941. Having suffered a lot under Mao Zedong's tyranny, he moved to Hong Kong in 1979. Since then he has worked until his retirement as a media analyst and translator at the American Consulate General in Hong Kong, a translator/interpreter for Baker & McKenzie, the translation manager of his own translation firm and chief editor at iSinoLaw China law website, which has joined forces with Thomson Reuters and appeared in Westlaw China's platform since November 2011.

Most people regard Tiananmen Protests as a failure. Chan, on the contrary, regards it as a major factor that facilitates the success of China's reform and peaceful coup d'état that enable China's miraculous rise. However, he is not surprised at the lack of response to his controversial description of Tiananmen Protesters' achievements. After all, it was an event more than two decades ago that few people still remember.

He knows his book will not bring about the rehabilitation of Tiananmen protesters in the near future. What he wants is to leave a book of history to let posterity know historical truth so that the Chinese Communist Party cannot distort it.

US well-known politician and diplomat Henry Kissinger's *On China* rouses the author's concern that people outside China even such a great diplomat as Kissinger lacks knowledge and understanding about China due to cultural differences and the mystery of Chinese politics.

To enable people to better understand China, Chan began to write blogs about China. In two years, he has made much viewed his microblogs on China at wordpress.com (tiananmenstremendousachievements.wordpress.com), LinkedIn under the name of Kai Yee CHAN and twitter at twitter.com/chankaiyee2.

In addition, he is a major contributor to chinadailymail.com, a well accepted website on China and has been invited by LinkedIn to contribute his posts.

The author's greatest concern is the possibility of the reemergence of a despot like Mao Zedong when China has surpassed the United States in military strength. Mao was only able to bring disasters to China due to China's weakness in Mao Era. The new despot may bring disasters to the whole world if he emerges after China has surpassed the U.S.

Previously, there is no urgency as China lags far behind the U.S. militarily. However, due to Obama's erroneous strategy and diplomacy and Chinese leader's ability to rally people around him to make joint efforts, China is winning in its arms race with the U.S.

That forces the author to write this book to enable people outside China to know the true situation that if the US does not switch to correct strategy and diplomacy now, China will defeat the U.S.